Kellogg
on Biotechnology

Kellogg
on Biotechnology
Thriving through Integration

EDITED BY

Alicia Löffler

Northwestern
University Press
Evanston, Illinois

Kellogg
School of Management
Evanston, Illinois

KOGAN
PAGE
London,
United Kingdom

Publisher's note
Every possible effort has been made to ensure that the information contained in this book is accurate at the time of going to press, and the publishers and authors cannot accept responsibility for any errors or omissions, however caused. No responsibility for loss or damage occasioned to any person acting, or refraining from action, as a result of the material in this publication can be accepted by the editor, the publisher, or any of the authors.

Northwestern University Press
Evanston, Illinois 60208-4170
ISBN 0-8101-2227-8

Kogan Page Limited
London, United Kingdom
ISBN 0 7494 4398 7

Printed in the United States

10 9 8 7 6 5 4 3 2 1

Copyright 2005 by Kellogg Center for Biotechnology, J. L. Kellogg School of Management. First published in 2005 in the United States by Northwestern University Press. First published in 2005 in Great Britain by Kogan Page Limited. All rights reserved. The right of Alicia Löffler to be identified as the editor of this work has been asserted by her in accordance with the Copyright, Designs and Patents Act 1988.

British Library Cataloguing in Publication Data

A CIP record for this book is available from the British Library.

Library of Congress Cataloging-in-Publication Data

    Kellogg on biotechnology : thriving through integration / edited by Alicia Löffler.
      p. ; cm.
    Includes bibliographical references and index.
    ISBN 0-8101-2227-8 (pbk : alk. paper)
    1. Biotechnology. 2. Biotechnology industries.
        [DNLM: 1. Biomedical Technology. 2. Biotechnology—economics. 3. Biotechnology—organization & administration. 4. Pharmaceutical. W 82 K29 2005] I. Löffler, Alicia. II. Kellogg Center for Biotechnology.
    TP248.2.K45 2005
    660.6—dc22

                    2004029045

The paper used in this publication meets the minimum requirements of the American National Standard for Information Sciences—Permanence of Paper for Printed Library Materials, ANSI Z39.48-1992.

CONTENTS

TABLES AND FIGURES

TABLES

FIGURES

PREFACE

We are drawn to this industry by the seductive allure
of achieving something meaningful—often expressed
in units of human life and well being. Instead of being
daunted by the complexity, or improbability of the
task, we are attracted to it. The intellectual challenge is
no small part of the satisfaction.
                          —*Richard Pops, chief executive*
                          officer of Alkermes, BIO CEO
                          Conference, February 25, 2004

B iotechnology is science-driven, ambiguous, risky, highly innovative, and ethically, socially, and politically charged. This complex and incredibly fast-moving field has forged a new future for
nearly every aspect of modern life, from the food we eat to the way
we approach diseases, clean the environment, and defend countries
from terrorism. Concomitantly, biotechnology is profoundly affecting business, education, and political priorities. Today, it is incon-
ceivable to think of a college graduate or business leader who
does not have some fluency in the language of biotechnology.
Biotechnology issues—from stem cell research to biodefense—have
also become centerpieces of political campaigns.

The term *biotechnology* is ambiguous and has an evolving definition. It does not encompass a single technology or industry.
Rather biotechnology is the convergence of innovative technologies
(biological/physical/computational) used to develop products and
processes that improve life in areas such as therapeutics, diagnostics,

medical devices, food, agriculture, and energy. Innovation in life science is the common denominator of biotechnology.

In fact, biotechnology's uniqueness comes not only from its prolific flow of innovations but also from the distinctive process by which it reaches its innovations. Biotechnology has redefined the process of innovation. Before the 1980s, knowledge creation was a linear, sequential process. The government provided funds to academia to conduct basic research, which was then turned over to the commercial sector for the development of innovations and commercial applications that benefited the general public. But since the 1980s, biotechnology has transformed this process by integrating academia, government, industry, and the public; knowledge and innovations are now emerging at the intersection of all these institutions. Consequently, basic research and applied research are no longer clearly distinct; indeed, they are often indistinguishable. This paradigm change is creating important policy challenges, particularly in the intellectual property domain. Consider, for example, GTC, an Australian company that focuses on the noncoding information in DNA (deoxyribonucleic acid) and has filed for patents to use information in the entire noncoding region of the genome. Is this applied research or basic research? And if intellectual property rights are granted to GTC, how will that affect the creation of basic research in this area in the future?

Integration is therefore the key word in innovation, which is one of the many reasons why biotech companies tend to cluster together around universities. In biotechnology, value lies in the integrated networks rather than in individual entities.

As the chapters in this book will make clear, developing a biotechnology product carries an unparalleled risk. A therapeutic product, for example, takes an average of twelve years to develop and an $800 million investment but has less than a 15 percent chance of reaching the market. In addition, biotechnology businesses are scrutinized vigilantly by regulatory agencies, which makes the process frustratingly sluggish.

Perhaps no other industry incites as much passion as biotechnology because of the polarized views on the social and political

Security Matters

While on board, please remember:

- Be aware of your surroundings.
- Keep your personal items secure and in close proximity. Laptop computers, PDA devices, portable music players, digital cameras, etc., are easy targets for pickpockets.
- Do not approach or pet police dogs.
- Report any suspicious activity or unattended luggage by notifying Amtrak Police and Security, personnel or by calling 1-800-331-0008.

Please feel free to consult a member of the on-board crew if you have any questions or concerns about this security program or if you need assistance. We thank you for your cooperation in helping keep the rail system safe and secure.

SEE SOMETHING suspicious or unusual?
SAY SOMETHING! Contact Amtrak Police and Security at 1-800-331-0008. Or Call 911.

Transportation
Security
Administration
www.tsa.gov

AMTRAK.
Amtrak.com

Operation ALERTS

Allied Law Enforcement for Rail and Transit Security

Amtrak® Police and Security, Transportation Security Administration (TSA) officials, and more than 100 police departments across 13 states and Washington, D.C., have mobilized today for Operation ALERTS (Allied Law Enforcement for Rail and Transit Security) — a joint, coordinated and synchronized rail security operation throughout the northeastern United States.

This train station is one of nearly 150 railway stations between Fredericksburg, Virginia and Essex Junction, Vermont involved in the operation.

Today's security deployment is NOT in response to any particular threat, but rather is part of an ongoing proactive approach to expand counterterrorism and incident response capabilities and enhance deterrence across Northeast Corridor railway systems.

During today's operation, passengers may notice enhanced security measures, including any of the following in stations or aboard trains:

- Uniformed police officers
- Uniformed TSA security officers
- Random passenger and carry-on baggage screening
- K-9 units
- Checked baggage screening
- On-board security checks
- Identification checks

Transportation
Security
Administration
www.tsa.gov

AMTRAK®

AmtrakCom

implications of the revolutionary technologies involved. For instance, genetically modified crops are considered by some to be a solution for world hunger, whereas others contend they are contaminating the food source. Similarly, some believe stem cell research saves lives, but others argue that it destroys life. The relationship between biotechnology and the public has its roots in the 1975 landmark Asilomar Conference, called to discuss the implications of recombinant DNA technology. Since that conference, the public has had an influential voice in shaping the biotechnology industry agenda, not only as customer and investor but also as moral gatekeeper.

Leaders in Washington realize that for U.S. biotech businesses to be competitive, the government must constantly review the policies promoting innovation in biotechnology while at the same time ensuring the ethics of the science behind the products developed in this field. In fact, we could argue that the government should take a much more active role in strengthening regulations and ethics and aligning them with the need for innovation and profit.

Often, in the absence of a quick response from governmental policymakers, leaders in the biotech industry create regulatory mechanisms for ethical self-policing. They do so not necessarily because they are humane (most of them are) but because they understand that the industry's future depends on public acceptance. Such actions, however, are quick and temporary fixes. The government will have to become more dynamic and responsive in meeting the industry's needs.

Leadership in biotechnology demands an uncanny ability to integrate skills and knowledge in science, business, politics, and ethics with the capacity to translate ideas into practice. Biotechnology requires a seamless transition across the diverse stakeholders' worlds to result in significant and permanent advances. Those responsible for planning business strategies in the biotech field need to respond to complex, disparate forces—from regulatory and political to technical and scientific—and yet never forget that the purpose of a business is to create profits.

FOCUS OF THIS BOOK

The biotechnology revolution has many stakeholders: the scientist who generates knowledge in a field where the pace is maddeningly fast and obsolescence is sudden; the technologist who turns an idea into a reality and is alert to the replacement of one promising technology with an even more promising one; the venture capitalist who provides funds based on market moods rather than fundamentals; individuals in the health care, agricultural, chemical, and pharmaceutical industries, all of which have been revolutionized and are now barely recognizable; the law professionals and ethicists who have to grapple with issues unimaginable just a few years ago; the public policymaker who struggles with complex issues and now must make decisions quickly or risk being left behind as other societies adopt a different course; the manager consumed by the increasing financial pressures of bringing a product to market who still needs to respond to the overwhelming complexity of the scientific, public, and political environments; and finally, the society and individuals in the midst of it all.

It is, of course, impossible for one book to address all these constituencies, but the chapters that follow provide examples from the biotechnology industry that are relevant to the biotech manager in the human therapeutic field. They offer an integrated view on the hurdles and strategies involved in reaching profitability in a complex technical, political, regulatory, and social environment. The unspoken message in these examples is that there should be a seamless integration across the many disciplines to further progress.

BIOTECH, "PHARMA," LIFE SCIENCE, AND MEDICAL DEVICES

In the early 1980s, after the advent of genetic engineering, biotechnology companies were recognized as those working on large molecules (protein therapeutics), whereas pharmaceutical ("Pharma") companies focused exclusively on small chemical entities. That division no longer exists. Today, biotechs and pharmaceutical companies both work with large and small molecules. For example, Genentech (a biotech company based in South San Francisco, California) is working on small molecules, and Abbott Pharmaceuticals (a Pharma company in Abbott Park, Illinois) is developing large molecules.

Biotechnology focuses on innovation in the life sciences, but innovation is not dependent on the size of a company. In fact, a high percentage of small life science companies are shamelessly uninnovative and at best just recycle value. Conversely, some flagship biotechs such as Amgen and Genentech, which are larger than traditional pharmaceutical companies, continue to generate an impressive flow of innovations in spite of their large size.

Innovation may be measured in many ways: output of patents, research and development (R&D) productivity, and the like. However, investors simplify the concept by rewarding perceived innovation with growth expectations, which then translate into high price/earnings (P/E) values. In this book, we often use the term *Pharma* to refer to large companies with industry average growth expectations that were founded as traditional chemical companies.

Although this book is devoted primarily to biotechnology in the human therapeutics area, any discussion of biotech should also address the medical device, diagnostic, industrial, and agricultural sectors. These sectors are becoming highly innovative and vibrant and are poised to deliver breakthrough products in the next few years. To whet the appetite of the reader, we have dedicated one chapter to medical devices.

THE SOURCE OF THE RESEARCH

The basis for this book is a collection of essays written by students in the Biotech Ventures course at the Kellogg School of Management. This course was first envisioned by Professor Mohanbir Sawhney in 1997 as a way of keeping up with the speed, intensity, and complexity of emerging technologies. Since then, the course has come to typify the spirit of Kellogg: students and faculty working in partnership to truly understand the latest business developments.

Students designed the syllabus for the Biotech Ventures course, including research trips to San Francisco, San Diego, and Boston, and they conducted secondary research that was followed by interviews with executives and industry experts. As the student mentor, I guided, reorganized, edited, and converted the essay content into chapters developing the flow of the book.

ORGANIZATION OF THE BOOK

The book is organized in three parts in order to mimic the growth of a biotech firm, from development of an innovative technology through to product commercialization: (I) Emerging Technologies: The Fuel of Biotechnology, (II) The Business of Science: Enhancing the Value of the Innovations through the Perfect Business Model, and (III) The Home Stretch: Commercializing and Capturing the Value of the Innovation. The parts are not meant to be comprehensive; rather, they highlight particular examples that illustrate the dynamics of the industry. Each part includes a short introduction to help put the examples in context.

PART I—EMERGING TECHNOLOGIES: THE FUEL OF BIOTECHNOLOGY

Part I describes three innovations that are bound to change the practice of medicine: stem cell technology, nanotechnology, and pharmacogenomics. All face significant hurdles that will have to be overcome before they can be widely adopted. Chapters 1 through 3 discuss these hurdles and identify some possible solutions.

Chapter 1, "Promises and Challenges of the U.S. Stem Cell Industry," describes the potential of the stem cell industry to revolutionize the treatment of diseases and conditions ranging from cancer to epilepsy and analyzes the importance of the timing of such developments, the full impact of government regulations, and how firms can address the incendiary ethical questions that come into play. Chapter 2, "Pharmacogenomics: Overcoming the Hurdles to Adoption," discusses some of the reasons for the reluctance of many pharmaceutical companies to invest in pharmacogenomics (the use of genetics to optimize drug discovery and development) and speculates on the strategies most likely to allow pharmacogenomics to succeed in the marketplace. Chapter 3, "Nanobiotechnology: Applications and Commercialization Strategies," addresses the challenges involved in integrating a nanotechnology innovation in biotechnology applications and provides a framework for managers to use in overcoming these challenges.

PART II—THE BUSINESS OF SCIENCE: ENHANCING THE VALUE OF THE INNOVATIONS THROUGH THE PERFECT BUSINESS MODEL

Revolutionary technologies do not necessarily produce good products, and good products do not always make good business. Like all emerging technologies, biotechnology firms go through periods of experimentation before adopting a business model that will best enhance the value of their technologies. Part II gives a short background on the three models usually employed by the biotechnology industry: platform, integrated, and mixed. Additionally, it looks at the issue of sustainability, focusing first on a particular model (platforms) and then on a tool used to augment survivability (mergers and acquisitions [M&A]).

Chapter 4, "Sustaining Platforms," investigates the platform model and discusses strategies to achieve long-term profitability. Chapter 5, "Mergers and Acquisitions as a Strategic Alternative in Biotech," presents a view on how acquisitions can be used as an alternative tool to increase the sustainability of a biotech firm.

PART III—THE HOME STRETCH: COMMERCIALIZING AND CAPTURING THE VALUE OF THE INNOVATION

Part III deals with some of the issues that biotechnology companies encounter as they move from science-driven to commercial-driven organizations. Particularly, it focuses on strategies for scaling-up manufacturing, marketing, and reimbursement functions. Because of space limitations, we do not address the complex issue of clinical trials, which is certainly important enough to deserve a book of its own.

Chapter 6, "Biologics Manufacturing: The Make or Buy Decision," details the options for outsourcing the manufacturing function and identifies some key tactical issues and future trends in the field. Chapter 7, "Pharming Factories," considers the possibility of manufacturing proteins in animals and plants and the obstacles that need to be overcome to make pharming a viable alternative to traditional manufacturing. Chapter 8, "The Role of Marketing," discusses the landscape of marketing in biotechnology and recommends how best to incorporate and optimize marketing in drug development. Chapter 9, "To DTC or Not to DTC: Direct-to-Consumer Marketing in Medical Devices," explores DTC marketing as a way to increase the marketing efficiencies of life science products. Part III concludes with Chapter 10, "The Forgotten Issue: Reimbursement in Biotechnology," which looks at the reimbursement strategies (or lack thereof) in biotechnology companies and suggests a framework for companies to use in considering the issue early in the development process.

Alicia Löffler

PART I

EMERGING TECHNOLOGIES: THE FUEL OF BIOTECHNOLOGY

Innovation is the fuel of the biotechnology industry. Biotech innovations arise from many fields (including biology, physics, computation, and engineering), but they all share a common goal—the development of products or processes designed to improve our health, environment, or agricultural resources.

Clearly, not all innovations make good products, and not all good products build profitable companies. Innovations are particularly difficult to assess because of the problems inherent in estimating all possible applications, market size, and time to market. Although quantitative assessments (net present values, risk-adjusted present values, options) provide us with concrete numbers, they also present us with an incomplete picture of the potential value of the technology. By using quantitative assessments alone, we will likely be overlooking the most intriguing and revolutionizing biotechnologies.

A more accurate assessment is possible when qualitative factors are also considered, such as inventive potential (whether the technology is revolutionary or evolutionary), applicability (whether the technology has a clear commercial application), and nonmarket environments (whether the

1

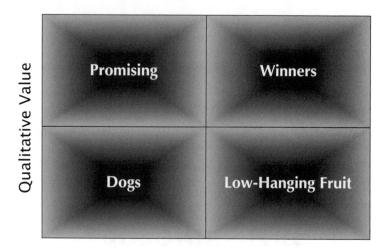

Figure I.A   Technology Assessment Quadrant

public, regulatory, and policy environments are ready to deal with the innovation).

Once qualitative and quantitative measurements are integrated, a quadrant emerges with which the different technologies can be compared and categorized as "winners," "promising," "dogs," and "low-hanging fruit," depending on their relative qualitative versus quantitative scores (Figure I.A).

In Part I of this volume, we qualitatively describe three revolutionary emerging technologies: stem cell technology, nanobiotechnology, and pharmacogenomics. Each is bound to play a dominant role in the personalized medicine market—the development of individualized therapies and cures. We do not attempt to exhaustively review the intricacies of the science (a few chapters are obviously not enough!); rather, we try to develop an instinct about what to look for in estimating the potential of a given technology.

Personalized medicine is an increasingly important market, particularly in light of the aging baby-boomer population; thus, we carefully considered the unique characteristics of this segment of the population. Baby boomers are highly educated and demand individualized treatments and cures for their diseases. Individualized health care and permanent cures are typified by prophylactic treatments, implantable devices, and stem cell regenerative medicine.

However, personalized medicine lies a decade or two in the future. It is expected that, in the near future, we will undergo a period of correlational medicine, in which innovations will be focused on establishing an association between diagnostics and monitoring of diseases with the new genomic discoveries. These correlations will increase our genetic understanding of the diseases and will pave the way for individualized therapies. Two key technologies that will play a critical role in advancing correlational medicine are nanobiotechnology and pharmacogenomics, which we discuss in the following chapters.

Chapter 1

PROMISES AND CHALLENGES OF THE U.S. STEM CELL INDUSTRY

Whit Alexander and Shail Thaker

Few technologies spark as much interest, hope, and controversy as stem cell technology. Many people have strong opinions about the morality of stem cell research, but few have a strong understanding of the science and its potential. In the highly charged atmosphere surrounding this technology, the United States has established federal restrictions that may be hindering the development of the domestic stem cell industry, leaving investors confused about the commercial feasibility of this technology.

In this chapter, we present a review of stem cell research and its potential applications. We then turn to the challenges that face the stem cell industry in the United States, including moral, legal, and political challenges. Finally, we analyze the implications for investors.

STEM CELL RESEARCH

Since their discovery in the early 1900s, stem cells (cells that have the ability to continuously divide and differentiate into specific cells/tissues) have captured the imagination of scientists. Interest intensified, however, in 1998, when Professor James Thomson at the University of Wisconsin isolated and grew stem cells derived from human embryos (Thomson et al. 1998). Soon thereafter, researchers from Johns Hopkins University achieved similar results with human germ cells (or cells from the gonadal ridge of aborted fetuses) (Gearhart 1998). These advances impelled a wave of stem cell research around the world, focusing on three areas: human development, birth defects, and therapeutics.

TYPES OF STEM CELLS

There are essentially two kinds of stem cells—embryonic and adult. Embryonic stem cells (ESCs) are harvested from the very early blastocyst stage of a human fertilized egg and are described as totipotent (able to develop into any body cell type, including placental tissue). Adult stem cells are found in developed tissue, such as bone marrow cells. They can be pluripotent (able to give rise to any type of cell in the body except those needed to develop a fetus) or multipotent (able to give rise to a small number of specific cell types, such as bone or muscle tissue). Hematopoietic stem cells—blood stem cells that can develop into several types of blood cells but no other types of cells—are one kind of multipotent stem cells.

A central dogma in the scientific community has been that at the end of embryo development, cells become "terminally differentiated"—that is, they are permanently committed to a specific function. However, recent research has shown that this state may not be as "terminal" as previously thought. Both adult blood cells and muscle cells have recently been induced to become undifferentiated (Hughes 2001). In addition, stem cells have been found in adult tissues, although these stem cells were initially thought to have neither the plasticity nor the longevity of their embryonic counterparts.

PROPERTIES OF STEM CELLS

Stem cells have two unique properties. First, they have *longevity*. They divide and replicate under laboratory conditions for long periods of time without differentiating until induced to do so. Second, they have *plasticity*. They are able to differentiate into different types of specialized cells, such as cardiac muscle or pancreatic cells.

Uncertainty exists over the genetic and/or environmental factors that determine the plasticity of cells. In vitro studies (studies outside the body) of embryonic stem cells are providing an opportunity to increase understanding of the role of biochemicals produced in the normal cellular environment that induce stem cells to differentiate, to migrate to sites needing repair, and to assimilate into tissues (Schuldiner 2000). However, much work is still under way.

SOURCES OF STEM CELLS

The most ready source of embryonic stem cells for research and therapeutic use is early-stage embryos created in vitro (through nuclear transfer), with the embryos then harvested at the blastocyst phase. Controversy exists regarding the use of this source of embryonic stem cells (see the later discussion on moral/ethical challenges to stem cell research). A less controversial source of embryonic stem cells, though one also not completely free from controversy, is embryos produced for the in vitro fertilization process.

Research with embryonic stem cells gives extraordinary hope for the treatment and cure of diseases, but researchers have a number of important hurdles to surmount, including ethical and technical concerns. The ethical concern, discussed in more detail later in this chapter, is that we might be destroying life by deriving cells from embryos, and the technical concern is that embryonic stem cells might be allogenic to the patient and prone to rejection.

Adult stem cells provide solutions to both of these problems. Despite their potential to have reduced plasticity relative to embryonic stem cells and despite the difficulty of harvesting sufficient amounts, adult stem cells may present a good alternative for medical applications.

The most common sources of adult stem cells are blood, bone marrow, and adipose tissue. Stem cells normally circulate in the blood in very small quantities and can be collected from the blood through a small catheter inserted into a patient's vein. Blood cell growth factors called cytokines can be administered to patients to produce a substantial increase in the number of circulating blood stem cells for collection. This process is referred to as stem cell mobilization. Two cytokines approved by the Food and Drug Administration that are commonly used for stem cell mobilization are Neupogen® by Amgen and Leukine™ by Berlex.

The collection of stem cells from bone marrow has been safely performed since the 1970s. Bone marrow harvesting is typically done in the operating room. The donor receives general anesthesia, and a large needle is inserted into the bone marrow cavity through a sterilized area of the lower back. Bone marrow is then aspirated. The needle is inserted multiple times, with a typical bone marrow harvest taking about two hours and resulting in about 1 liter of bone marrow being removed. The major side effect is discomfort at the site of the bone marrow harvest.

Adipose, or fat, tissue is an exciting source of adult stem cells due to the high number of stem cells it contains. Fat tissue containing stem cells can be extracted from the body through liposuction. Some stem cells derived from fat tissue have been found to have multipotent activity (Zuk et al. 2001) and have been reprogrammed into other cell types, including muscle, cartilage, and bone cells (Gearhart 1998; Weiss 2001).

POTENTIAL THERAPEUTIC APPLICATIONS

One of the most fascinating medical applications of stem cell research is in regenerative medicine, which uses stem cells to generate tissue that can repair failing organs. With the potential to change medical practice in both treating and curing disease, the applications of regenerative medicine span a wide array of debilitating diseases. The following paragraphs describe a few of these promising applications, including heart disease, diabetes, and neurological disorders.

Heart Disease

Stem cell technology provides hope for the cure of heart disease. Congestive heart failure affects 5 million people in the United States. With this disease, heart muscle is damaged and functional tissue is replaced with nonfunctional scar tissue, reducing the ability of the heart muscle to contract. Geron Corporation has generated functioning human cardiomyocytes (heart muscle cells) in culture from embryonic stem cells that can potentially replace the scar tissue. Some academic communities are, off the record, hoping to have some form of treatment derived from embryonic stem cells available for cardiac muscle damage by 2005 or 2006.

Insulin-Dependent Diabetes Mellitus

Another potential therapeutic application for stem cells is in the treatment of insulin-dependent diabetes mellitus, a disease that affects approximately 1.4 million Americans. The daily insulin injections prescribed to control the body's blood sugar levels have profound effects on the patient's quality of life. Embryonic stem cells induced in vitro to become insulin-secreting cells and transferred to diabetic animals have restored normal glucose balance within a week (Dor et al. 2004).

Neurological Disorders

Perhaps the most astounding application of stem cells is in the treatment and cure of neurological disorders. Researchers recently transplanted embryonic stem cells into the brains of mice and found that the stem cells differentiated into dopaminergic neurons, restoring partial (80 percent) neural function in the rat and mouse model of Parkinson's disease (Björklund et al. 2002; Fallon et al. 2000). Similarly promising results have been observed in rats with spinal cord injuries (Ramón-Cueto et al. 2000).

Recent Advances in Stem Cell Research

Two recent developments have the potential to radically alter the direction of research in the development of stem cell therapies. First, the *Proceedings of the National Academy of Science* (Mezey et al. 2003) reported the results of a study on autopsies of women who had received bone marrow transplants from male donors. The researchers found neuronal cells containing the Y chromosome in the brain, implying that stem cells from the bone marrow transplant migrated from bone to brain. The implications are that stem cell therapies do not need to be administered locally and that they could be relatively easily delivered because they will migrate to the "correct" target.

Second, Catherine Verfaillie and her colleagues at the University of Minnesota (Westphal 2002) found that adult stem cells derived from bone marrow were capable of differentiating into every type of human tissue, eliminating the possibility of immune rejection—a property previously believed to be characteristic only of embryonic stem cells. An implication of this research is that therapeutic cloning—the cloning of human embryos to obtain stem cells that the body will accept—would no longer be necessary.

MORAL/ETHICAL DILEMMAS OF STEM CELL RESEARCH AND DEVELOPMENT

With stem cells, we were beginning to have all of the issues we were worried about in bioethics in the same room: consent, women as research subjects, sex, IVF [in vitro fertilization] clinics, animals, abortion, germ line intervention, and aging.

—*American Journal of Bioethics*
(Zoloth 2002)

The excitement about stem cell research and its potential therapeutic applications is clearly evidenced by the large investments into research that have been made by corporations, governments, and universities around the globe. Tempering this exuberance, however, are the myriad ethical, legal, and political challenges that face this field of research. The eventual resolution of these conflicts will determine the success of the research and potentially the face of medicine in the future.

Ethical challenges are not new to medicine and medical research. Indeed, the earliest code of medical ethics, the Hippocratic Oath, sits at the heart of the discussion on embryonic stem cell research. According to the Hippocratic Oath, those who swear by it are bound to do no harm. With embryonic stem cell research, the question arises of whether it is "harm" to destroy an embryo created for research or a surplus embryo that was developed for in vitro fertilization to obtain stem cells that will be used to heal.

As with many other issues surrounding medical research, divergent opinions abound. Even those beliefs based in religious conviction can be starkly opposed. For example, whereas the Catholic Church extends its arguments against abortion to propose banning research on surplus embryos from in vitro fertilization clinics, some interpretations state that Jewish law mandates research with such un-implanted embryos to "save life" (Zoloth 2002).

Additionally, stem cell research raises the concern that humans are "playing God." This concern is particularly relevant when discussing the cloning of human embryos for research purposes. There is also the possibility of inadvertent germ-line (reproductive cells) manipulation, even with the use of adult stem cells. Germ-line manipulation would result in the genetic modification of the offspring and would have a permanent impact on the human species.

A discussion of ethics in stem cell research cannot, however, be limited to the research arena. As therapies are developed and commercialized, society will have to consider the ethical implications of not only the science but also the management of the corporations that bring such treatments to market. Several of these issues are discussed in more detail in the following sections.

TISSUE SOURCE

The source of tissues used to obtain stem cells is one of the most incendiary topics surrounding stem cell research. The President's Council on Bioethics suggested three primary and recurring points of contention with regard to the issue of tissue sources: the moral status of human embryos, complicity, and the "alternative of adult stem cells" (Outka 2002). The first of these points, moral status, refers to the inherent worth of something from a moral standpoint (rather than an economic, technical, or other standpoint) (Brock 2002). The key question is, When does an embryo become a human?

The political Right argues that human embryos are human at the moment of conception and that "the human individual . . . must always be treated as an end in himself or herself, not merely as a means to other ends" (Doerflinger 1999). To destroy a fetus in the course of research, then, is simply to take the life of an innocent. This unambiguous assertion holds that all embryonic stem cell research is morally wrong, regardless of potential benefits.

The political Left, in contrast, argues that becoming human requires more than the potential to develop into a human child. This argument is based in the belief that nonimplanted embryos are "too rudimentary in structure or development" and thus that "no moral duties are owed to embryos by virtue of their present status" (Robertson 1999). Following this logic, it is morally acceptable to destroy surplus or cloned embryos for the greater benefit of society.

Other voices in the debate attempt to distinguish between types of embryos. Specifically, some people hold that before implantation, the embryo is not "an individualized human entity with the settled inherent potential to become a human person" and that "use for certain kinds of research can be justified" (Farley 2000). For many people, however, this distinction falls short of allowing research on aborted embryos or fetuses.

The second point raised by the President's Council on Bioethics, complicity, relates to the question of who is responsible for embryo destruction. Are scientists who conduct research on embryonic tissue complicit in embryo destruction? One argument is that regardless of the source of the embryo (creation for research, surplus from in vitro fertilization clinics, or other sources), researchers are complicit because the act of doing research destroys a living embryo. Additionally, such destruction is assumed and undertaken as a part of the research protocol. For those who take this view, the "rightness" of conducting research depends largely on moral status: "*Either* we should stop opposing the creation and destruction of embryos for research purposes only . . . *or* we should oppose not only the creation and destruction of embryos for research purposes, but the research on spare embryos from in vitro fertilization clinics too" (Outka 2002). Another view of complicity makes a distinction between the destruction of surplus embryos and the destruction of embryos created specifically for research. In the former case, the decision to destroy the embryo has already been made by somebody else, not the researcher.

The third point of contention noted by the President's Council involves "the alternative of adult stem cells." Some contend that therapies derived from adult stem cells are not sufficient to advance research and innovations. Consequently, they argue, research using stem cells from *all* sources should go forward.

FUNDING AND ACCOUNTABILITY

Given these unresolved ethical controversies and society's apparent acceptance of stem cell research, the ethics of the science behind stem cell research is now being taken up in the context of the public's role in funding and oversight. To the extent that federal funding of research in any way suggests society-wide convictions, it is no wonder the debate has reached such a fevered pitch. If one holds embryonic stem cell research to be immoral, then providing tax dollars to support such "wholesale killing" must also be immoral.

As the commercialization of treatments becomes a reality, any ethical debate will have to include accountability (MacDonald 2002). Issues relating to moral status and complicity are weighed in thousands of company boardrooms by senior managers and directors who have an eye on profits. Biotech companies face increasing pressure to assure good science, strong financial performance, and high moral grounds.

Some of the managers' dilemmas relate to the science itself—for example, how to select appropriate cell lines. Should the decisions be dictated by financial prospects? Should ethics also play a role? At the expense of potentially saving thousands or millions of lives, should managers opt for less controversial cell lines? Other areas in which managers have to balance profits with maximizing social well-being include patenting and the affordability of treatments.

Several patents related to stem cells have already been granted in the United States and abroad. The National Institutes of Health (NIH) has encouraged companies to take "generous" stances on licensing to allow the continuation of research (Freire 2001), yet some managers argue that current federal regulations are blocking development and would likely opt for more restrictive patents. Finally, as treatments are brought to market, companies will naturally attempt to find reliable (and less expensive) sources of basic raw materials. Given the controversies surrounding cloning and the creation of embryos for research, it is likely that other sources will appear. Indeed, a market for "disposable" human tissue has grown, allowing biotech companies and researchers to secure a supply of fetal tissue (Jiminez 1999). Managers face the ethical dilemma of determining how to secure a reliable source of materials for treatment while satisfying ethical considerations of moral status.

LEGAL ISSUES

A report issued by the NIH in 2000 stated that "congressional prohibition does not prohibit the funding of research utilizing human pluripotent stem cells because they are not embryos" (National Institutes of Health 2001). However, the report also stated that appropriations law (P.L. 105-277, section 511,112 STAT, 2681-386) prohibits funds "for the creation of a human embryo or embryos for research purposes; or research in which human embryos are destroyed, discarded or knowingly subjected to risk of injury or death." The problem, of course, is that human embryonic stem cells are derived from early embryos. The proposed NIH answer is to permit funding for pluripotent stem cell research while denying funding for deriving stem cells from embryos. As stated in the report, government funds can be used "only if the cells were derived from early human embryos that were created for the purposes of infertility treatment and were in excess of clinical need of the individuals seeking such treatment." This makes quality embryonic stem cell lines a very valuable asset.

THE POLITICAL LANDSCAPE

FUNDING FOR RESEARCH

Faced with substantial moral dilemmas, the U.S. government took a cautious position toward stem cell research. In 1995, a federal statute prohibited the use of tax money for research on human embryos (Allen 2001). In 2000, the Clinton White House skirted these restrictions by allowing funding for stem cell research provided the destruction of embryos was not done with federal money. For those who take an absolutist position in terms of complicity, this policy was unacceptable. In the face of such concerns, the Bush administration took the government's position on stem cell research decidedly "rightward," though not far enough for some staunchly pro-life members of the Republican Party (Allen 2001). In a televised address in August 2001, President George Bush announced new guidelines for embryonic stem cell research that allowed NIH funding only with the following conditions:

> Removal of cells from the embryo must have been initiated before August 9, 2001 . . . and the embryo from which the stem cell line was derived must no longer have had the possibility of developing further as a human being. The embryo must have been created for reproductive purposes but no longer be needed for them. Informed consent must have been obtained from the parent(s) for the donation of the embryo, and no financial inducements for donation are allowed. (NIH 2003)

Congress has historically taken a much stronger position against stem cell research. Following Clonaid's 2002 announcement that it had successfully cloned a human, the House and Senate moved to ban human cloning. Soon after, however, in 2003, bills were introduced to take steps to protect forms of stem cell research (House Resolution 081.IH and Senate Resolution 303.IS [U.S. Government 2003]). Against this backdrop, some leaders of the Republican majority, including Senate Majority Leader Bill Frist, advocated providing funding for embryonic research, suggesting that any future legislation would be mindful of the value that such research promises (Frist 2001). The execution of federal funding and regulation decisions falls largely on the NIH.

VOICES IN THE DEBATE

In a society driven by public opinion, a number of forces must be recognized in a stem cell discussion. One of these forces is the country's aging population, a group that will be disproportionately impacted by many of the proposed benefits of stem cell research. Organizations such as the AARP (formerly the American Association of Retired Persons) will likely become more vocal in their support and lobbying for strengthened federal stem cell efforts. Another powerful force is the politics of celebrity in the United States. Within the context of stem cell research, celebrities such as Michael J. Fox, whose health might benefit from stem cell research, have become vocal champions for all forms of stem cell research. Working in contrast to these forces are a number of religious groups and the political Right. What remains to be seen is whether lawmakers as well as other political players will make a distinction between scientific promise and their religious convictions.

COMMERCIAL FEASIBILITY OF STEM CELL TECHNOLOGY

THE INDUSTRY IN THE UNITED STATES

With the tremendous promise of stem cell technology, it is not hard to understand the significant amount of public and private investment in the technology. In the United States, more than twenty companies are solely dedicated to developing therapies based on stem cell research. With the possibility of products that could revolutionize the treatment of diseases and conditions ranging from cancer to epilepsy, the potential for profit is staggering. The challenges lie in the timing of such developments and, as discussed earlier, the impact of government regulations and the management of the ethical questions that come into play.

Four key U.S. companies conducting material embryonic stem cell research are Geron in Menlo Park, California; PPL Therapeutics in Blacksburg, Virginia; Bresagen in Athens, Georgia; and Advanced Cell Technology in Boston. Geron, the first mover into the sector, is considered the dominant company of the group. The key U.S. company conducting adult stem cell research is ViaCell (based in Boston). Other companies involved in elements of adult stem cell research are Layton Biosciences, in Sunnyvale, California, and Osiris Therapeutics, in Baltimore. None of these companies is profitable. In fact, the leader, Geron, had a net loss of $33.9 million in 2002. Indeed, the U.S. market for stem cells is nonexistent. In 2001, the total sales by stem cell companies in the United States were a mere $51.7 million. Assuming that cell therapies are able to capture 10 percent of therapeutic expenditure on neurodegenerative diseases and on cardiac muscle treatment, the market could, in a ten-year time frame, represent a market of over $10 billion. Should the technology deliver on the promise of early research, even conservative observers believe this market will generate billions of dollars in annual sales.

Given the political landscape and the technology's potential, managers of companies in the stem cell industry clearly should develop strategies of market diversification and private funding. Equally important to their companies' success, they should have a political and public strategy in place. For example, these managers should be actively monitoring and participating in upcoming stem cell legislation. However, because the stem cell issue is already entrenched in the legislative phase (i.e., legislative bodies are already involved, and legislation is extant), the degree of freedom available for managers is limited (see Figure 1.1).

Recognizing how organized the pro- and anti-stem-cell lobbies are, any attempt by scientists to pass favorable legislation or block unfavorable legislation may require the breakup of the opposing coalition. The pro-stem-cell lobby works through trade organizations (such as the Biotechnology Industry Organization, or BIO)

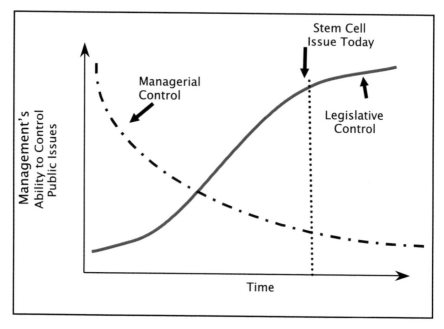

Figure 1.1   Life Cycle of Public-Charged Issues: Legislative versus Company's Management Control

Source: Adapted from Diermeir 2002.

and through patient and stakeholder groups (such as AARP, neurological disease patient support groups, and the Patients' Coalition for Urgent Research). The anti-stem-cell lobby works through traditionally conservative and religious groups.

The biotech industry's economic and health benefit arguments supporting stem cell research have had some success, notably with statements by pro-life Senators Orrin Hatch and Bill Frist displaying their support for embryonic stem cell research. Also, the support of high-profile individuals such as Michael J. Fox has increased the positive perception of this research in the public eye. However, these voices have not been sufficiently effective in breaking up the opposing coalition made up of religious leaders and groups (including Richard Doerflinger of the National Conference of Catholic Bishops and the Christian Legal Society); some doctors, scientists, and groups to which they belong (such as former Surgeon General

C. Everett Koop and the Coalition of Americans for Research Ethics); and pro-life/family advocate groups (such as the Center for Bioethics and Human Dignity).

Arguments from the anti-stem-cell lobby are based on three core ideas: (1) embryonic stem cell research is "tainted" because of intrinsic links to cloning, (2) the destruction of embryos is morally wrong, and (3) the United States can maintain a position of scientific dominance as well as deliver on the promise of cell therapy by adult stem cell techniques alone.

Should all attempts of advocates of stem cell research at lobbying messages to the anti-stem-cell interest groups fail (economic arguments to conservatives, pressure from patients in religious communities, and so forth), managers of stem cell concerns and scientists in the field will likely look abroad to invest and to set up operations in a more conducive regulatory environment.

REGULATORY ENVIRONMENT ABROAD

Policies on stem cell research vary greatly from nation to nation. Britain is leading the drive for acceptance. In early 2001, the U.K. Parliament approved the amendment of the 1990 Human Fertilisation and Embryology Act governing research on human embryos, which allows the use of embryos up to fourteen days old for research on the derivation and potential of human stem cells. Under this legislation, embryos may be legally created for stem cell research by somatic cell nuclear replacement (SCNR), although the 2001 Reproductive Cloning Act made reproductive cloning illegal. In February 2002, the Lords committee (Upper House), chaired by the bishop of Oxford, concluded that the potential benefits of embryonic stem cell research justified the use of early human embryos. In September 2002, the Medical Research Council and Biotechnology and Biological Sciences Research Council announced that a contract had been awarded to the National Institute for Biological Standards and Control to establish a national repository of stem cell lines. Following these landmark rulings, the United Kingdom has become the largest public funder of stem cell research in the world.

In Israel, a national bioethics committee in 2001 approved the derivation of embryonic stem cells and research into therapeutic cloning. The report of that committee, the Bioethics Advisory Committee of the Israel Academy of Sciences and Humanities, obtained the force of law when the national science funding agency agreed to follow its recommendations.

In Australia, an agreement was reached in April 2002 between the prime minister and the leaders of the nation's eight states and territories to enact national legislation allowing embryonic stem cell research using "spare" in vitro fertilization embryos that would otherwise be destroyed but prohibiting reproductive cloning—a policy not dissimilar to that followed in the United States.

In Canada, legislation was introduced giving legal force to current guidelines issued by the Canadian Institutes of Health Research permitting the use of public funding for research on stem cells derived from surplus in vitro fertilization embryos up to fourteen days old but prohibiting therapeutic cloning.

Japan has also allowed stem cell research. In March 2000, the bioethics committee of Japan's Science and Technology Council (Tokyo) approved research involving the use of embryonic stem cells (Saegusa 2000). Subsequent guidelines instituted in 2001 stipulated that embryonic cells used in research would be taken only from embryos made for fertility treatment that would otherwise be discarded. This stipulation is in line with Japan's history of conservative regulation on medical research. Research on cloning humans or creating sperm and ova is strictly banned in Japan.

The European Union (excluding the United Kingdom) has also taken a progressive stand regarding stem cell research. In November 2000, the European Group on Ethics in Science and Technology (EGE), which advises the European Commission, expressed the opinion that in countries where research on human embryos is already legal, there are no additional ethical barriers to the use of such embryos for stem cell research. However, in its view, the creation of embryos specifically for this purpose was not justified given the availability of "surplus" in vitro fertilization embryos, and it was deemed premature to allow somatic cell nuclear replacement. The European Parliament, in a narrow vote in September

2000 (European Parliament 2000), condemned the United Kingdom's proposals to allow somatic cell nuclear replacement in stem cell research, and in August 2001, the European Parliament's Temporary Committee on Human Genetics produced a report that (after heavy amendment of an originally more moderate text) recommended a ban on the use of European Union funding for research involving somatic cell nuclear replacement and a ban on the creation of embryos specifically for research purposes. However, the temporary committee's report was rejected by the European Parliament in November 2001. Reproductive cloning is condemned by the Council of Europe in an additional protocol to the European Convention on Human Rights and Dignity of the Human Being with Regard to the Application of Biology and Medicine, and Article 18 of the European Convention prohibits the creation of human embryos for research purposes. The convention and protocol have both been ratified, so that signatories (about half of the member states of the Council of Europe but not including, for example, the United Kingdom, France, Germany, Italy, or the Netherlands) are obliged to incorporate them into their national legislation by 2006. In Germany, Parliament voted in January 2002 to allow research on imported human embryonic stem cell lines that had been created before January 20, 2002. These cell lines can be used only for research projects approved by a new regulatory body.

In China, research on embryonic tissue is generally banned, according to the Chinese Health Ministry. However, the study of stem cells drawn from the umbilical cord and afterbirth is permitted. Chinese institutions are very aggressive in many areas of genetic research, and regulation is somewhat lax.

Given that other countries including Israel and the United Kingdom have made stem cell research a scientific and commercial imperative and "are pouring massive amounts of government money into it" (Mitchell 2002), it is not surprising that many fear the United States is facing an imminent "brain drain." Since the well-known American scientist Roger Pederson moved to Cambridge University from the University of California in 2001, many U.S.-based scientists have repeated his transatlantic journey. As stated by Linda Powers, managing director of Toucan Capital

(Bethesda, Maryland), "The migration of research to outside the United States would be a big blow from a business and economic standpoint" because treatments and technologies resulting from stem cells "are going to yield very big economic returns" (Mitchell 2002). Most investors conservatively predict the market will be in the billions of dollars.

WHAT DOES ALL OF THIS MEAN TO INVESTORS?

In the United States, legal and policy issues related to stem cell research probably will not have a tremendous public health impact. Although federal restrictions that address concerns from the Right about embryonic research have hindered stem cell research, they have not limited the use of any potential therapies. However, a crucial policy question in the United States is whether the strength of the "conservative" lobby will prevent the country from developing a competitive advantage in biotechnology. One must also ask how or how much policy restrictions will hinder the development of the domestic biotech industry and whether the United States will be forced to import future therapies. With these oppositional forces in place, will the United States stop being the leader in life science innovation? As discussed earlier, many nations of the world are clearly moving forward and seizing the opportunity. In moving ahead early with funding, these countries may be able to secure a substantial stake in the industry and perhaps an insurmountable lead over U.S. interests.

Investors (both businesses and individuals) in the stem cell industry looking to maximize returns clearly should consider the odds for a successful therapeutic product. Given the basic advantages (including plasticity) of embryonic stem cells over adult sources, an investment that does not incorporate an opportunity to capitalize on such embryonic stem cell research would be risky. Indeed, researchers working with embryonic stem cells are likely to derive therapies before researchers working with adult sources of stem cells. Although U.S. corporations are currently participating in embryonic stem cell research (and despite recent advances with adult sources), overseas stem cell companies are poised to be the first to bring cell therapies to market. Undoubtedly, U.S. domestic firms will also bring products to market, but given the current policies in the United States that limit research, the prudent investor might be wise to look elsewhere.

REFERENCES

Allen, C. 2001. "Bush Cell-Out." *National Review Online,* August 13. Available at http://www.nationalreview.com/comment/comment-allen 081301.shtml.

Björklund, L., R. Sánchez-Pernaute, S. Chung, T. Andersson, I. Ching Chen, K. McNaught, A. Brownell, B. Jenkins, C. Wahlestedt, K. Kim, and Ole Isacson. 2002. "Embryonic Stem Cells Develop into Functional Dopaminergic Neurons after Transplantation in a Parkinson Rat Model." *Proceedings of the National Academy of Science* 99: 2344–2349.

Brock, D. W. 2002. "Bioethics: Messing with Mother Nature," review of *Our Posthuman Future,* by Francis Fukuyama. *American Scientist* On Line, September. Available at http://www.americanscientist.org.

Diermeir, D. 2002. "Strategic Management in Non-Market Environments." Lecture presented at the Kellogg School of Management. Fall 2002.

Doerflinger, R. M. 1999. "The Ethics of Funding Embryonic Stem Cell Research: A Catholic Viewpoint." Kennedy Institute of Ethics *Journal* 9, no. 2: 137–150.

Dor, Y., J. Brown, O. Martinez, and Douglas A. Melton. 2004. "Adult Pancreatic β-Cells Are Formed by Self-Duplication Rather Than Stem-Cell Differentiation." *Nature* 429: 41–46.

European Parliament. 2000. "Text of European Parliament Resolution Adopted on September 7, 2000." Available at http://www.europarl .eu.int/1gc2000resolution#B5-0710/2000.

Fallon, J., S. Reid, R. Kinyamu, I. Opole, R. Opole, J. Baratta, M. Korc, T. Endo, A. Duong, G. Nguyen, M. Karkehabadhi, D. Twardzik, and S. Loughlin. 2000. "In Vivo Induction of Massive Proliferation, Directed Migration, and Differentiation of Neural Cells in the Adult Mammalian Brain." *Proceedings of the National Academy of Sciences USA* 97 (December 19): 14686–14691.

Farley, M. 2000. "Roman Catholic Views on Research Involving Human Embryonic Stem Cells." Presented at the forty-second National Bioethics Advisory Commission meeting, Bethesda, Md., July 2000.

Freire, M. 2001. "Statement of National Institutes of Health before Senate Appropriations Subcommittee on Labor, Health and Human Services, Education and Related Agencies." Senate testimony, August 1. Available at http://www.stemcells.nih.gov/policy/statements/080101/freire.asp.

Frist, W. 2001. "Frist Announces Support for Stem Cell Research." Senator Bill Frist Press release, July 15. Available at http://frist.senate.gov/index.cfm?FuseAction=PressReleases.Home. Accessed on November 2, 2004.

Gearhart, J. John. 1998. "Potential for Human Embryonic Stem Cells." *Science* 282: 1061–1062.

Hughes, S. 2001. "Muscle Development: Reversal of the Differentiated State." *Current Biology Review* 20, no. 6: R237–R239.

Jiminez, M. 1999. "U.S. Investigates Traffic in Foetal Parts." *National Post* (Canada), November 27. Available at http://www.all.org/news/natpost1.htm. Accessed June 2004.

MacDonald, C. 2002. "Stem Cell Ethics and the Forgotten Corporate Context." *American Journal of Bioethics* 2, no. 1: 54.

Mezey, Eva, Sharon Key, Georgia Vogelsans, Ildiko Szalayoua, G. David Lange, and Barbara Crain. 2003. "Transplanted Bone Marrow Generates New Neurons in Human Brains." *Proceedings of the National Academy of Science* 100: 1364–1369.

Mitchell, S. 2002. "U.S. Stem Cell Policy Deters Investors." United Press International. Available at http://www.upi.com/view.cfm?storyID=20021101-053230-6291r.

National Institutes of Health. 2001. *Guidelines for Research Involving Human Pluripotent Stem Cells.* Available at http://www.hhs.gov/news/press/2001pres/01fsstemcell.html.

———. 2003. "NIH Stem Cell Backgrounder." *Stem Cell Information, National Institutes of Health.* Available at http://www.nih.gov/news/backgrounders/stemcellbackgrounder.htm. Accessed on March 17.

Outka, G. 2002. "The Ethics of Stem Cell Research." Meeting of the President's Council on Bioethics, April. Available at http://www.bioethics.gov/topics/stemcells_index.html.

Ramón-Cueto, A., M. I. Cordero, F. F. Santos-Benito, and J. Avila. 2000. "Functional Recovery of Paraplegic Rats and Motor Axon Regeneration in Their Spinal Cords by Olfactory Ensheathing Cells." *Neuron* 25 (February): 425–435.

Robertson, J. A. 1999. "Ethics and Policy in Embryonic Stem Cell Research." *Kennedy Institute of Ethics Journal* 9, no. 2: 109–136.

Saegusa, A. 2000. "Japan Okays Stem Cells." *Nature Biotechnology* 18 (March): 246.

Schuldiner, M. 2000. "Effects of Eight Growth Factors on the Differentiation of Cells Derived from Human Embryonic Stem Cells." *Proceedings of the National Academy of Sciences USA* 97: 11307–11312.

Thomson, J. A., J. Itskovitz-Eldor, S. S. Shapiro, M. A. Waknitz, J. J. Swiergiel, V. S. Marshall, and J. M. Jones. 1998. "Embryonic Stem Cell Lines Derived from Human Blastocysts." *Science* 282: 1145–1147.

U.S. Government. 2003. House Resolution 081.IH and Senate Resolution 303.IS. Available at http://frwebgate.access.gpo.gov/cgi-bin/multidb .cgi. Accessed June 2004.

Weiss, Rick. 2001. "Human Fat May Provide Stem Cells." *Washington Post,* April 10, A01.

Westphal, S. 2002. "Ultimate Stem Cell Discovered." January 23. Available at http://www.newscientist.com/news/print.jsp?id=99991826. Accessed on November 2, 2004.

Zoloth, L. 2002. "Stem Cell Research: A Target Article Collection, Part 1— Jordan's Banks, A View from the First Years of Human Embryonic Stem Cell Research." *American Journal of Bioethics* 2, no. 1: 3–11.

Zuk, P. A., M. Zhu, H. Mizuno, J. Huang, W. Futrell, A. Katz, P. Benhaim, H. P. Lorenz, and M. H. Hedrick. 2001. "Multilineage Cells from Human Adipose Tissue: Implications for Cell-Based Therapies." *Tissue Engineering* 7, no. 2: 211–228.

Chapter 2

PHARMACOGENOMICS: OVERCOMING THE HURDLES TO ADOPTION

Janette Chung, Peggy Mathias, and Rebecca Wildman

An outgrowth of the Human Genome Project, pharmacogenomics (the use of genetics to optimize drug discovery and development) is the next research frontier, and significant investment in the industry is anticipated. For now, however, smaller peripheral companies and biotechnology companies are the ones focusing on pharmacogenomics, and "Big Pharma" remains on the sidelines waiting for certain validation before it allocates substantial levels of investments.

Investments are guided not only by market concerns but also by nonmarket (patient, regulatory, and ethical) considerations. Economic factors—measures of profitability and predicted potential profitability—traditionally drive pharmaceutical investment decisions, but patient demand and controversial ethical issues are the tipping points that will influence Big Pharma's decision to embrace this technology. The regulatory context acts as the enabler—serving as an impetus (or deterrent) to full adoption. The Food and Drug Administration (FDA) position on pharmacogenomics will encourage entry into and competition in the field and reward companies for innovation and progress. Similarly, rules on orphan drug status and patent protection must be addressed to further encourage innovation.

WHAT IS PHARMACOGENOMICS?

Pharmacogenomics is an umbrella term that includes the use of genetics to optimize drug discovery and development. In this chapter, we use the term broadly to refer to tailor-made drugs or personalized medicine—that is developing the right drug for the right people. Personalized medicine is the marriage of functional genomics and molecular pharmacology. Pharmacogenomics seeks to find and decipher correlations between patients' genotypes (genetic profiles) and their therapeutic responses.

Pharmacogenomics uses these correlations to discover new and highly effective therapies tailored to specific genetic makeups. The process involves identifying genes and their protein offspring as potential drug targets and then understanding the variations of the genes. For example, multiple forms of one gene might prompt the creation of drugs suited for each specific gene variation. Interest and funding in pharmacogenomics is largely fueled by growing evidence that an individual's genetic profile is and will continue to be the key predictor of how effective particular therapies are.

Given the complexities of tying genes to diseases, some companies are trying a different approach and are attempting to connect genes directly with existing drugs. For example, Genaissance Pharmaceuticals, located in New Haven, Connecticut, is studying people with different haplotypes (alternative forms of part of a gene complex) to determine why they respond differently to treatments for what appear to be the same symptoms. Genaissance's primary project is studying statins, which are drugs used to regulate cholesterol levels in the blood. A variety of patients responded differently to each of the four drugs in the trial, and the company is beginning to understand the differences between the haplotypes for these patients. The hope is that these data can be used to predict patient responses to various drugs, thereby minimizing side effects and increasing the efficacy of individual treatments.

Currently, large pharmaceutical companies operate under what has become known as the blockbuster model, which can lead to situations in which a minority of products drives the majority of revenue. Blockbuster drugs are typically characterized as drugs with over $1 billion in revenue per year. In addition to creating a significant amount of revenue, they have product appeal beyond the initial target (long life-cycle management), and the demand for them exceeds capacity, typically due to unmet or newly addressed therapeutic needs. They are first-in-class drugs or improved versions of already effective drugs. Pharmaceutical companies adopt the blockbuster model because of high drug development costs that, combined with research and development (R&D) costs, are estimated to exceed $800 million per successful drug (Geiger 2003).[1] In contrast to the blockbuster model of the large pharmaceuticals, the nature of pharmacogenomics is specialized and denotes a more fragmented model tailored to segments of the population. This dichotomy is the most powerful force impeding the pursuit of pharmacogenomics by Big Pharma.

In addition, although many believe in the scientific value of pharmacogenomics, the industry still has many skeptics. Dr. Ted Love, president and corporate executive officer (CEO) of Nuvelo, Inc., contends that pharmacogenomics is just too complex. He has stated, "We don't even fully understand what genes cause diabetes, so how are we going to figure out what causes some people to react toxically to some pharmaceutical products?"(Love 2003). Others argue that although pharmacogenomics has incredible potential, we are too impatient in our desire to realize its promise. Dr. Heiner Dreismann, president of Roche Molecular Diagnostics, contends that, as with most new ideas, the concept of pharmacogenomics was too widely hyped at the outset and is now being too strongly critiqued during the establishment phase; he predicts it will eventually reach a level of acceptance somewhere in the middle of these two extremes. He believes strongly that pharmacogenomics will be an important factor in the future of medicine but cautions that the long discovery and development times in the industry mean we must be patient before the benefits of pharmacogenomics are realized (Dreismann 2003).

Two potential general business strategies may be appropriate for the emergence of pharmacogenomics. The first is what we will call a toxicity screening strategy, in which a drug developer seeks to identify and exclude those patients for whom a drug would be toxic.

The second strategy is a niche efficacy strategy and is employed when a drug company deliberately targets specific patients for whom the drug will be effective while excluding all others. We hypothesize that the former may be more attractive to pharmaceutical companies because it offers the continued ability to reach and serve broad targets, whereas the efficacy strategy is more likely to be favored by peripheral pharmaceutical players or emerging biotechnology companies.

TOXICITY SCREENING STRATEGY

The genetic makeup of an individual can determine his or her reaction to a pharmaceutical, and the distribution of reactions among a group of individuals can be envisioned as the area under a bell curve. Although the vast majority of patients react acceptably to a product, the reactions of a few will fall under each tail of the curve—at one end, a small number of patients will have a severe adverse response to the product, and at the other end, patients will have an unexpectedly positive response (Figure 2.1).

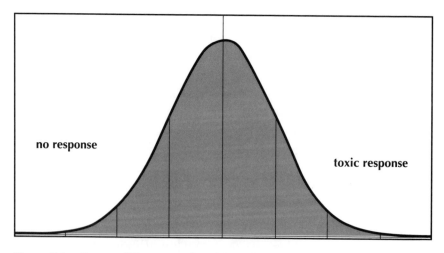

Figure 2.1   Range of Responses for a Typical Drug

Investment in pharmacogenomics allows a company to detect and understand the genetic profiles of people who are prone to adverse effects and consequently eliminate them from the patient population. In this case, multiple parties win because the pharmaceutical company can release the drug to the market and recoup the heavy R&D investment and patients who are not predisposed to "toxicity" can benefit from the drug. Physicians also stand to benefit, for their confidence in prescribing the drug should rise as the ability to assess patient reaction increases.

This strategy would seem to be a lucrative one for a pharmaceutical company. However, despite the ability to exclude so-called toxic patient populations and the potential to vary the dosage of existing medicines to service genetic segments and thus extend the life cycle of the drug, we have not yet seen mass adoption of pharmacogenomics by large pharmaceutical companies. A primary reason for this is the lack of understanding about genetics (Kreeger 2001), and the science is made even more complex by findings that prove many diseases are caused by a combination of genes rather than just one gene. A second reason is that the toxicity model demands increasing investments in discovering, developing, and securing FDA approval for both a diagnostic *and* a therapeutic. Moreover, this model requires many big pharmaceutical companies to change their organizational structures. In many Pharmas, the therapeutic and diagnostic arms operate as independent business units competing within the organization for limited resources. Pharmacogenomics requires these units to work together and make compromises in their profit structures.

An additional reason for the lack of adoption is that the requirement for diagnostics is only feasible for some diseases. For example, patients may be willing to take a diagnostic before having access to a cancer drug but may not necessarily be willing to do so for a cholesterol-lowering product. Finally, the large supply of niche players and biotechnology firms specializing in pharmacogenomics allows pharmaceutical companies to take a bystander role. Pharmas are making small in-house investments in pharmacogenomics to keep current with science and other industry leaders while watching for success in the arena and positioning themselves to acquire strong biotechnology or niche player targets.

Minicase: Seldane[tm] (terfenadine)— Toxicity Model

Seldane, a prescription medication for allergies commonly referred to as hay fever, was marketed by Hoechst Marion Roussel for ten years and was popular due to its ability to relieve symptoms without causing drowsiness. However, in a small number of people, Seldane was associated with a potentially fatal heart condition, ventricular arrhythmia. Eight deaths were attributed to Seldane use before the FDA withdrew its approval of the drug in 1997.

An understanding of genetics explains Seldane's toxicity effects and raises questions about how this profile could have been addressed. Seldane is metabolized by cytochromes CYP2D6 and CYP3A4. However, 10 percent of the population has no CYP2D6, and therefore all metabolism responsibilities are shunted to CYP3A4. Several common drugs are inhibitors of CYP3A4, including some antibiotic and antifungal drugs (such as erythromycin and ketoconozole). For patients who relied solely on CYP3A4 for metabolism, the coadministration of these inhibitor drugs caused Seldane levels to build up in the body, which resulted in arrhythmias.

Could an understanding of genetic profiles together with the targeted sale of Seldane to individuals who enjoyed the benefits of the drug without being genetically at risk for adverse side effects have resulted in the continued sale of this blockbuster product?

COMMERCIAL STRATEGY

Biotechnology companies and niche pharmaceutical companies can take advantage of the trend toward fragmented markets in an efficacy model. These companies can look at the successful orphan drug market—which addresses diseases affecting less than two hundred thousand people—to see the potential revenue generation in serving smaller niche segments.[2] In this model, R&D teams consider segment patient populations by genotype in the hopes of isolating different variations. Once variations are located, drug discovery efforts can be focused on one or many of these variations to create tailored drugs for each. The success of the drug Herceptin, from Genentech, shows the promise of this strategy. Herceptin is aimed at a relatively small population, specifically, the 35 percent of breast cancer sufferers whose *HER2* gene is over-expressed. Estimates show that Herceptin has been prescribed to 15 percent of those women with breast cancer caused by the *HER2* gene. In developing this drug and using testing (in partnership with a Danish diagnostic test maker, DAKO), Genentech was able to screen for its clinical trials and enroll smaller targeted patient populations; thus, fewer people and fewer trials (and much lower costs) were required to achieve statistical significance (Genentech 2003a). Although the patient population is extremely small compared to the mass markets reached by blockbuster drugs and despite the fact that it was only launched in U.S. markets in 1998, Herceptin had annual sales of around $188 million in 2000, $244 million in 2001, and $278 million in 2002 (Genentech 2003b). Success in such markets depends on having small targets and a small reach coupled with high prices (which can be achieved due to the higher efficacy and lower cost of trials for targeted pharmacogenomic drugs).

This strategy, which has brought success to many smaller companies, is not without its challenges. The segmentation of the market and advances in technology bring an influx of fierce competition from other niche and biotechnology companies. Also, ethical issues are an extremely dominant force in this arena. The toxicity model may be shielded from some of these issues, for the goal there is to maximize drug benefits for the largest community possible. But the efficacy model will likely face ethical issues in regard to the selection of the gene subtype or the patient population segment. These issues will be especially controversial because genotypes vary across race, gender, and/or other demographic lines. Along with these ethical issues, we will also see continuing concerns about security and the value of personal information, much like we have seen with other technological revolutions (such as the Internet). The public is likely to raise questions about the privacy, storage, and confidentiality of genotype information, as well as the unwarranted use of personal pharmacogenomics data.

Four primary factors will influence the adoption of pharmacogenomics by large pharmaceutical companies: economic justification, patient demand, resolution of ethical dilemmas, and regulatory backing.

MARKET JUSTIFICATION

The pharmacogenomics vision includes three key goals: to increase efficacy and reduce risk to patients, to develop diagnostics that impact therapeutic decisions and improve patient care, and to improve clinical development outcomes. These goals must all be accomplished while allowing for attractive economic returns. Pharmacogenomic products will remain unattractive to Big Pharma unless the economic impact of having a smaller target market is offset by either decreased development costs or premium pricing. Specifically, the attractiveness of pharmacogenomics to pharmaceutical companies relies on the following factors:

- Lowered discovery and development costs through more-targeted research efforts
- Development of diagnostic tests that are accurate and economically justified
- Justification for premium pricing
- Sufficient market size
- Payer support through adequate reimbursement
- Surmountable marketing issues
- Resolvable public and ethical issues

LOWERED DISCOVERY AND DEVELOPMENT COSTS

The most widespread use of pharmacogenomics today is in the discovery phase. Expression arrays, genetic maps, and molecular methods for studying gene function and expression are routinely utilized, allowing companies to more efficiently identify promising leads.

But to date, pharmacogenomics has rarely been applied in the development phase. A primary problem in proving the statistical significance of drug efficacy is that the patient population in clinical trials is further subdivided by genetic factors. The average number of patients receiving a drug before a new drug application (NDA) is less than four thousand, and if only a portion of those patients demonstrate the applicable genetic factors, the relevant sample size decreases and statistical significance is difficult to attain. Thus, larger trials may be required, with the potential to increase the cost of development significantly.

However, by targeting a predefined portion of the population and performing clinical trials for only those patients who fit that genotype, more-targeted therapies can be tested using smaller trials to statistically prove significance. In addition, smaller trials can decrease approval time, which is a critical factor: as Dick Brewer, former president and CEO of Scios, Inc., has stated, "The single most important thing in drug development, aside from approval, is time" (Brewer 2003). Although smaller trials can decrease clinical trial costs and speed time to approval, it also limits the potential market for the approved compound.

To increase the likelihood of cost-effectively attaining a commercially relevant result from pharmacogenomic research, many companies are focusing on the genes that impact drug action rather than the genes that determine disease. They choose this approach because the vast majority of diseases are influenced by a combination of genetic factors, whereas the kinetics, safety, and efficacy of drugs appear to be more often determined by significant single gene effects. For example, most of the common mutations in genes linked to cancer account for no more than 5 percent of the disease incidence. Segmenting genetic factors based on a drug's effects may provide more consistent and statistically significant results.

ECONOMICALLY JUSTIFIED DIAGNOSTIC TESTS

Pharmacogenomic therapies require the development, approval, and marketing of not only a therapeutic agent but also the associated diagnostic, entailing additional costs. Successful pharmacogenomic products must be able to leap the additional economic hurdle these costs introduce before being economically justifiable.

PREMIUM PRICING

The benefits of a more-targeted and effective product with fewer side effects must be of sufficient magnitude to allow for premium pricing, as compared to the pricing of conventional therapies. Consequently, pharmacogenomic products are most likely to be effective in therapeutic areas such as cancer treatment, where diseases are acute, adverse reactions are both frequent and significant, and the efficacy of drugs is high when specified. In these areas, payers are quite likely to reimburse very costly treatments and patients are not as concerned with pricing because the treatments often make the difference between life and death. Because pharmacogenomic products tend to be extremely efficacious in the selected population, adoption rates are arguably higher. In addition, the application of pharmacogenomics to developing more-targeted versions of existing blockbusters and extending patents by devising more-targeted formulations is of particular interest to large pharmaceutical companies.

MARKET SIZE

Pharmacogenomic therapies are prescribed for patients who display certain genetic traits; therefore, these genetic factors must be common enough to create a sizable patient population. The requisite size is determined by the other economic characteristics of the product, which, combined with sales volume, will determine the profitability of the product. For example, although only approximately 25 percent of breast cancer sufferers overexpress the *HER2* gene and are therefore eligible for Herceptin, the price charged for the product justifies its continued sale from an economic standpoint.

REIMBURSEMENT

Managed-care organizations have strong incentives to control medical expenditures, and thus, they impose price controls and other reimbursement limitations that could have negative impacts on innovation in pharmacogenomics. But because pharmacogenomic therapies have relevance to only limited patient populations and because they tend to be more efficacious, the payers' risk is very limited; therefore, the payers are more willing to reimburse therapy costs. This ease of reimbursement will generally exist as long as pharmacogenomic drugs have low aggregate costs. However, the situation can change, as it did in the case of Herceptin. As the efficacy of Herceptin became more widely known and adoption rates increased, the aggregate costs of the drug no longer "flew under the radar," and reimbursement issues became more of a concern for the payers (Parks 2003).

As pharmacogenomic products become more common, difficult managed-care issues may arise. First, patients in genetic subgroups may be prone to certain diseases or may have only costly therapies available to them. The health insurance industry will be tempted to charge higher premiums for individuals in these subgroups, an action that most would consider discriminatory. Although personalized medicine will always face this issue, we believe legal protections and vocal consumer groups will continue to work to ensure equity for all segments of the population.

SURMOUNTABLE MARKETING ISSUES

Pharmacogenomic products introduce new factors into the cost of marketing. For example, marketing to a diseased population when only a portion of the patients are eligible for treatment raises issues regarding the cost of marketing. In addition, because the drugs treat only a subset of the patient population, sales forces must be effectively stratified. Salespeople must also be trained not only on the drug itself but also on the differences in drugs for specific gene types; consequently, a more knowledgeable and perhaps a more scientifically minded sales force will be required. Finally, additional efforts will have to be made to educate doctors about the use of a diagnostic-therapeutic combination.

For a pharmacogenomics product to be successful, these additional marketing costs must be minimized once the product is launched. Prior to launch, the decision to enter the market could be influenced by the potential strategies for marketing. As more companies begin to integrate the R&D teams and marketing development in truly cross-functional structures, these strategies must prove cost-effective enough for drug companies to take on the commercial risk of the drug.

As pharmaceutical companies are coping with rising costs and relying more heavily on the emergence of blockbusters and "megabrands" to recoup their investments, new consumer-focused messages have saturated the market in order to increase demand and awareness. Coupled with the fact that traditional marketing to physicians has become harder (both in terms of reaching doctors through direct office contact and in terms of stricter regulations on medical education), there has been a marked shift to direct-to-consumer (DTC) marketing by pharmaceutical companies. In 1997, the DTC floodgates were opened when the FDA allowed pharmaceutical companies to advertise prescription drugs using their specific names and the conditions they treat. Previously, the FDA had only allowed pharmaceutical companies to advertise new therapy options for diseases, without naming the particular drugs; alternatively, if a drug was named, the company could not mention the condition it treated ("Pharmaceutical Advertising and You" 2003). Since then, the public has seen a torrent of both television and print ads that educate the patient and employ a push strategy, whereby the consumers then ask their doctors about the drugs being promoted. Due to the early success of DTC advertising, spending in this category has increased from $1.8 billion in 1999 to $2.6 billion in 2001 (Matthews 2003).

Successful drugs advertised through DTC typically fall into one or more of the following categories (Findlay 2002):

1.   Lifestyle drugs, which provide cosmetic or lifestyle benefits that are valuable to consumers but do not necessarily impact long-term outcomes of serious illnesses.

2.   Prevention drugs, which are intended to decrease the likelihood that disease complications will occur.

3.   Chronic disease drugs, which affect the maintenance of the patients' long-term health. Consumers focus not only on the specific drug but also on how the drug fits into the treatment approach, including reactions to other medications, ease and frequency of dosing, and other therapies.

Drugs in the preceding categories have seen great success in marketing terms, in part due to their large patient populations and also because they treat diseases that are typically easy to self-diagnose (such as heartburn) or compelling enough to induce patients to seek treatment (such as erectile dysfunction or obesity). Pharmacogenomics runs directly contrary to these "criteria for success" for DTC marketing in that it attempts to attract targeted populations that require testing to determine if patients "qualify" for the drug therapy.

This fragmentation of the market may lead to a shift from the development of actual drugs to the development of diagnostic tests. The makers of diagnostic tests, who potentially stand to profit substantially from the advent of pharmacogenomics, will increase their partnerships with large drugmakers and copromote testing and drugs. Copromoting with diagnostic companies (or vertically integrating into diagnostics) may replace or offset large-scale DTC television and print campaigns and increase the yield in terms of user population, for if a patient is tested for a drug and "qualifies," the likelihood that he or she will take it increases substantially. Although we do not yet see mass adoption of pharmacogenomics, we do see clear trends, as evidenced in various partnerships. Consider, for example, the partnering between Biogen and Genaissance. Biogen's senior vice president of corporate development, John Palmer, agrees with his colleagues that pharmacogenomics is far off due to technological and scientific limitations, and he predicts that the environment will likely be fraught with privacy issues and concerns about the use of personal information. Nonetheless, Biogen has recently partnered with Genaissance, whose technology helps identify associations between a patient's genetic makeup and his or her drug response (Genaissance Pharmaceuticals 2002).

Vaughn Kailian, the former vice chairman of Millennium Pharmaceuticals, states that the pharmacogenomics industry is the "single greatest opportunity and the single greatest risk in our future." Though he, like most others in the industry, believes that advances in pharmacogenomics lie too far in the future to warrant the allocation of resources to analyze marketing trends, he also feels that mass adoption will occur as costs decrease (in clinical trials) and efficacy increases. He contends that the FDA will push for the increased safety and efficacy that pharmacogenomics bring; however, mass adoption will not occur until the costs of screening and development decrease. Kailian believes that the potential for pharmacogenomic products will be realized through marketing to educated consumers, and he agrees that conventional DTC is not the approach of choice for that effort. Rather, he predicts that the Internet will increasingly be used as a marketing vehicle and that activist groups will induce consumers to take a battery of tests to determine "cocktail" treatments.

Pharmacogenomic products must pass the same economic muster that all other therapeutics and diagnostics do; however, they must accomplish this while being handicapped with smaller target markets and additional costs. This situation effectively raises the bar for pharmacogenomics products, a fact that, to date, has dissuaded Big Pharma from making any significant investments in personalized medicine. We anticipate that the first wave of pharmacogenomic products will

- Address relatively large target markets
- Treat acute diseases whose current treatments have severe and frequent side effects
- Command premium pricing

- Have accurate diagnostic tests
- Not introduce new ethical and legal issues

RESOLVABLE PUBLIC ISSUES

Advocacy on behalf of patients has proven to be a powerful force in the regulation and development of drugs, and in the past, it has succeeded in driving new product introductions. For example, AIDS advocacy groups that emerged through the 1980s and 1990s amplified their voices to force change because of the desperate need for access to treatment. They increased public awareness with demonstrations and medical education programs, raised research funding, educated patients, and fought for coverage for diagnostic testing and treatments. They were so successful in their efforts that pharmaceuticals, health care providers, insurers, and government agencies eventually included the AIDS advocacy groups in their strategic planning. And by creating a tipping point involving regulatory, social, and reimbursement issues, these groups changed the pharmaceutical industry and compelled it to make a commitment to treating the disease. Similarly, breast cancer and Parkinson's advocates have modeled themselves after the AIDS advocacy groups, politicized their issues, and shaped the research direction of pharmaceuticals. Today, advocacy groups are poised to have a strong influence in pushing various parties to fulfill unmet medical needs through the adoption of pharmacogenomics in the years ahead.

In the development of pharmacogenomic therapies, patient advocacy groups may also play a crucial role in speeding up the clinical test process by providing access to patients with certain genetic profiles. Once again, Herceptin is a success story on how prescriptions contingent on the results of a diagnostic test can hasten the drug-approval process and improve drug efficacy. Before the drug was approved, the breast cancer advocacy groups helped Genentech amend the clinical trials to make them more appealing to patients; they also helped the company determine a fair way to provide the experimental medicine to eligible, seriously ill patients. Then, they worked with Genentech for regulatory guidelines, with advocates participating at investigator meetings. Finally, they cooperated with the company to raise awareness in targeted populations. By similarly developing partnerships with advocacy groups, other pharmaceutical companies can improve the development, approval, and adoption processes in the future.

The benefits of pharmacogenomics to patients seem obvious: more specifically targeted products provide more efficacious treatments with fewer side effects. However, the FDA's approach to pharmacogenomic products will have a significant impact on how widely and quickly these products are adopted. Although the FDA is currently taking a largely reactive position in regard to pharmacogenomics, it has issued two guiding documents that discuss the importance of genetic variations in studying the metabolism of drug candidates. These documents identify pharmacogenomics as one of several important modalities for studying metabolism. The effects of genetic polymorphism on dosage in achieving the safe and effective use of drugs is recognized in one document, indicating the FDA's increasing emphasis on pharmacogenomics (Rubenstein 1999).

APPROVAL OF PHARMACOGENETIC TESTS

Pharmacogenomics is characterized by the use of a pharmacogenetic test to identify patients with appropriate genetic profiles before prescribing the therapeutic drug. The FDA makes a distinction between pharmacogenetic tests that are manufactured and packaged as kits for sale to multiple laboratories (which are considered "diagnostic devices") and tests that are developed by individual laboratories and offered directly to patients (which are called "home brews"). The former are subject to FDA approval; the latter are not. Congress has heard testimony from experts in the field calling for greater regulation of the home brews to assure their safety and efficacy, and because therapies using these home brews are increasing, it is expected that the FDA will extend its authority to oversee these tests as well. In either case, coapproval of therapeutic drugs and pharmacogenetic tests by the FDA may deter the development of pharmacogenomics due to the added complexity and cost entailed in this additional approval hurdle.

To attract pharmaceutical companies to invest in pharmacogenomics, the FDA must provide more specific guidelines regarding pharmacogenetic test regulations. The Secretary's Advisory Committee on Genetic Testing (SACGT) suggested several parameters for developing a scheme for classifying pharmacogenetic tests and determining the level of scrutiny each deserves.[3] These parameters address:

1. The accuracy of genetic tests
2. The clinical usefulness of the tests
3. The issue of population-based screening
4. The risks of predictive tests versus diagnostic tests

In addition to clearer guidelines for pharmacogenetic tests, the FDA can encourage pharmaceutical companies to invest in pharmacogenomics by streamlining the coapproval of a therapeutic drug and the corresponding diagnostic test.

LABELING

Given the gatekeeping function of pharmacogenetic tests in the prescribing of drugs, reliable and accurate communication of the predictive values of test results becomes critical for both the toxicity model and the efficacy model described earlier in this chapter. First, pharmaceuticals need to make claims about the drug's safety and efficacy based on the genotypes of patients. These claims may become the basis for drug approval. Second, pharmaceuticals need to convey information on the drug dosage for each genotype and the required diagnostic test. This step may require new labeling and prescription guidelines.

Labeling also affects the acceptance of drugs by physicians. Currently, physicians can use drugs for purposes other than those approved by the FDA if the benefits to the patients can reasonably be expected to exceed the risks. There might be two "off-label" uses in the context of pharmacogenomics: the therapeutic drug might be prescribed to a patient who is not a good candidate for it, or it might be prescribed without first conducting a diagnostic pharmacogenetic test. These two off-label uses must be managed in order to provide guidelines for physicians and encourage the adoption of pharmacogenomics.

ORPHAN DRUGS

Pharmacogenomics will allow researchers to identify disease subsets because the response to a specific genotype can be identified. As a result, what might be considered a homogeneous disease could actually involve an aggregate of genotypes, each responding to a different therapy. It is possible, then, that pharmacogenomics will lead to the identification of more subsets of a disease population, thereby potentially allowing for preferential treatment under the Orphan Drug Act. Through this act, the government offers tax credits, research grants, exclusive marketing rights, and other valuable incentives to companies to encourage research on rare diseases. The FDA does not regulate prices for orphan drugs; pharmaceuticals are free to set their own. Since the inception of the act, 231 orphan drugs have been approved, facilitating treatment for an estimated 11 million patients in the United States (Haffner, Whitley, and Moses 2002). As a result, pharmacogenomics is expected to benefit from the Orphan Drug Act by identifying medically plausible subsets in a phenotypically similar disease group.

SOCIAL AND ETHICAL ISSUES

Pharmacogenomics is raising new issues in the ethical hotbed of biotechnology. Although the toxicity model tries to maximize the population with the inclusion of as many genotypes as possible, the efficacy model of personalized medicine involves targeting patients with specific genotypes, which raises important ethical challenges. For example, in 2005, Nitro Med (Lexington, Massachusetts) will launch the first "black" drug, Bidil, specifically targeted to treat heart failure in African Americans (Peerson 2004). Beyond the clear concerns of using skin color as a therapeutic category, the introduction of this drug also involves issues of fairness. How does one address the ethical problem of allowing only a portion of the diseased population to receive the therapeutic if there are no equally efficacious alternatives available to patients for whom the drug does not work? Other ethical questions are raised by the specificity and sensitivity of the diagnostic test. For instance, how does a company ensure that patients are not denied treatment based on the negative results of diagnostic tests?

Both models may also arouse fears about racism because, by nature, pharmacogenomics highlights the differences in genotype among individuals and populations. Certain types of genetic variations that are of importance in the metabolism of drugs are known to be more common in some ethnic groups than in others. If adverse responses are associated with a particular ethnic group, members of the group might suffer from stigmatization. Similarly, if one treatment serving an ethnic group comes to market more quickly than another, issues will certainly surface about placing greater "value" along racial lines. In diverse countries such as the United States, it will be difficult to ensure that therapies for all ethnic groups are fairly addressed.

Personalized medicine also prompts concerns about the security and privacy of a patient's pharmacogenetic information. Among those concerns, issues relating to informed consent and secondary information are key.

Patients will more readily accept pharmacogenomic testing if their rights to consent to the testing are fully protected. First, the information gathered in pharmacogenetic tests, like that in many diagnostic or clinical lab tests that are now routinely administered with at most minimal informed consent, should theoretically carry no major risk of psychosocial harm. Second, in determining what information should be collected from patients, the benefits of diagnostic information and the costs to privacy should be balanced.

Pharmacogenomic tests might also disclose sensitive secondary information about patients. Test results might, for example, be linked to a genetic disease that is of great interest to employers and insurers. Or pharmacogenetic information suggesting a patient will not respond to a particular drug might be considered by employers or insurers as an indication of an untreatable serious illness. In addition, pharmacogenetic tests might reveal sensitive information about family members, such as facts regarding paternity. As a result, the level of patients' confidence in pharmacogenomics—and thus the development efforts by pharmaceuticals—will be influenced by the regulation of privacy issues related to pharmacogenetic information.

SUMMARY

Pharmacogenomics tailors therapies to the genetic makeup of an individual and can therefore offer treatments that are more efficacious and have fewer side effects. Despite these benefits, personalized medicine has not been embraced by large pharmaceutical companies. We have suggested two models for pharmacogenomics—one that focuses on excluding patients for whom a given treatment would be toxic and one that tries to identify patients for whom the treatment would be effective. We believe that pharmaceutical companies will adopt pharmacogenomics under the toxicity model, as this model still includes a large portion of the patient population and yet avoids some of the difficult ethical issues associated with the efficacy model. In addition, we suggest that the first wave of successful pharmacogenomic products will be used in acute treatments for which current therapies have frequent and severe side effects; these products should also be good candidates for premium pricing. However, for pharmacogenomics to be truly embraced, the benefits of this technology must become more widely accepted even as the economic, public, regulatory, and ethical issues are addressed.

NOTES

1. Other sources cited include TAP Pharmaceuticals, Pharmaceutical Research Manufacturers of America (PhRMA), the FDA, and the Tufts Center for the Study of Drug Development.

2. The single most important criterion used to identify rare disease is a prevalence level below 7.5 affected individuals per 10,000 people. The threshold was included in the U.S. Orphan Drug Act of 1983 and currently approximates 200,000.

3. SACGT was chartered by Health and Human Services in 1998 and was superseded by the Secretary's Advisory Committee on Genetics, Health and Society in 2002.

REFERENCES

Brewer, Richard. 2003. Discussion with Kellogg Biotech Ventures class during visit to Scios, Inc., March 16.

Dreismann, H. 2003. Discussion with Kellogg Biotech Ventures class during visit to Roche Molecular Diagnostics, March 18.

Findlay, Steven. 2002. "DTC Advertising: Is It Helpful or Hurting?" National Institute for Health Care Management. Statement before the Federal Trade Commission Health Care Workshop, September 10. Available at http://www.ftc.gov/ogc/healthcare/findlay.pdf.

Geiger, Doug. 2003. "Role of NPD Marketing." TAP Pharmaceuticals presentation at the Kellogg School of Management, January.

Genaissance Pharmaceuticals. 2002. "Genaissance Pharmaceuticals and Biogen Sign Pharmacogenomics Collaboration." Press release, February 1.

Genentech, Inc. 2003a. SEC filings and Genentech announcement at J. P. Morgan conference, January 8.

————. 2003b. Roundtable discussion with Kellogg Biotech Ventures class during visit to Genentech, March 19.

Haffner, Marlene E., Janet Whitley, and Marie Moses. 2002. "Two Decades of Orphan Product Development." *Nature Reviews Drug Discovery* 1 (October 2002): 821–825.

Kreeger, Karen Young. 2001. "Scientific, Ethical Questions Temper Pharmacogenetics." *The Scientist* 15, no. 12: 32.

Love, T. 2003. Discussion with Kellogg Biotech Ventures class during visit to Nuvelo, Inc., March 17.

Matthews, M. 2003. "Who's Afraid of Pharmaceutical Advertising?" Institute for Policy Innovation Policy Report no. 155.

Parks, D. 2003. "Biotherapeutics and Managed Care." Presented at Kellogg Biotechnology Conference, Evanston, Ill., April 5.

Peerson, H. 2004. "First Black Drug Nears Approval." *Nature On Line,* July 23. Available at http://www.nature.com/news/2004/040719/full/040719-16.html.

"Pharmaceutical Advertising and You." 2003. Health Partners (Bloomington, Minn.) Report, March. Available at http://www.healthpartners.com/Menu/0,,3902,00.html#1.

Rubenstein, Ken. 1999. "Pharmacogenomics—Impact on Drug Discovery." Drug Market Development Reports, October. BioPortfolio, Frampton, Dorset, UK.

Chapter 3

NANOBIOTECHNOLOGY: APPLICATIONS AND COMMERCIALIZATION STRATEGIES

Badri Amurthur, Barry Grant, Markus Hildinger, Eric Hyllengren, Mohammed Ladha, Michael Masters, Daniel Shuffrin, and Jing Ye

In its purest definition, the term *nanotechnology* refers to a specific manufacturing technology in which larger molecular structures are built atom by atom (from the bottom up) in the nanometer (nm) scale (1 nanometer is 1 billionth of a meter). This concept was first introduced by physicist Richard Feynman in his 1959 lecture "There Is Plenty of Room at the Bottom." Feynman stated, "The principles of physics, as far as I can see, do not speak against the possibility of maneuvering things atom by atom." In this respect, nanotechnology represents a new frontier, in which phenomena and processes are different from those utilized in traditional macro- and microtechnologies. However, the concepts of nanomanufacturing, self-assembly, and bottom-up synthesis of larger molecular complexes are not completely new: nature itself relies on exactly those principles. Every living cell is composed of natural nanomachines such as proteins and nucleic acids that interact to produce tissues and organs. The goal of nanotechnology is to mimic and exploit

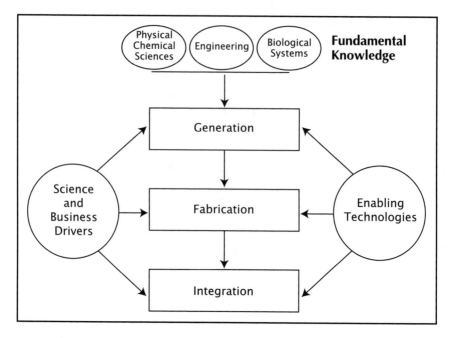

Figure 3.1   The Ecosystem in Nanotechnology Development

similar principles by gaining control of structures and processes at atomic, molecular, and supramolecular levels and to learn to efficiently manufacture and apply nanomachines.

Novel ideas are generated from the vast reservoir of fundamental knowledge that currently exists across several disciplines. These novel ideas are then translated into feasible technologies (Figure 3.1). Recent advances in nanotechnology (such as carbon nanotubes, molecular motors, DNA-based [deoxyriboneuclic acid-based] assemblies and computing, quantum dots, and molecular switches) represent just the first messengers of the revolution to come (Drexler 2000).

There are two different approaches to nanotechnology fabrication—the bottom-up approach and the top-down approach, as graphically represented in Figure 3.2. The bottom-up approach,

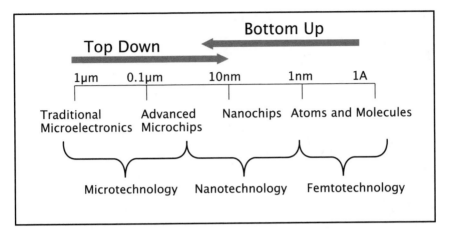

Figure 3.2   Overview of Technologies and Their Scale of Application

Note: A = angstrom; μm = micron; nm = nanometer.

or molecular nanotechnology, is used to build organic and inorganic structures atom by atom or molecule by molecule.

The top-down approach is an extension of microtechnology wherein selective etching is used to fabricate nanosystems. The top-down approach owes its origin to the growth and development of microelectro mechanical systems (MEMS). The field of MEMS, which was firmly established in the mid-1980s, has matured to a point where miniature motors can be mass-produced. MEMS is the confluence of semiconductor processing and mechanical engineering on a micrometer scale. Nanoelectro mechanical systems (NEMS) are the next step toward miniaturization in this field. With advancements in surface and bulk nanomachining techniques, NEMS can now be built with extremely small size (approximately 10 nanometers) and mass (10^{-18} grams).

Nanotechnology stands poised to have an enormous impact on biotechnology and medicine. The comparable size scale of the nanoparticles and biological materials, such as antibodies and proteins, facilitates their use for biological and medical applications.

APPLICATIONS

Two forces are driving applications in nanobiotechnology: the sheer force of the scientific advances that provide revolutionary enabling technologies and the tremendous need to surmount specific medical and pharmaceutical hurdles.

For example, scanning probe technology, including scanning tunneling microscopy and atomic force microscopy, has enabled powerful developments in the field of nanoscale technology. The power of these enabling techniques has provided the impetus for developing even more highly performing scanning probe tips, fabricated through microfabrication techniques. The development of a tip technology impacts the synthesis and the assembly processes themselves. Scanning probe technologies are used as the basis for materials that pattern and process at nanometer scales. Sophisticated in situ monitoring techniques provide a greater understanding of and control in the synthesis of nanostructure building blocks. Other important advancements include sensing and detection technology, separations technology, chemical analysis of nanoliter volumes, and nuclear magnetic resonance and enhanced computational infrastructure.

The need to find solutions to the current pharmaceutical and medical problems is propelling progress in nanobiotechnology. The inefficiency of the Pharma pipeline presents a challenge that has been difficult to overcome thus far. Currently, only one in a thousand target drugs ends up as a product, and it costs $800 million and takes twelve years to develop one drug. This sort of productivity should be unacceptable given our technical sophistication. Failures in the drug development process could be reduced by using better detection devices early in that process to identify drugable leads. This need presents a huge opportunity for the rise of nanotechnology.

Nanobiotech also offers the opportunity to treat and prevent diseases by taking totally novel approaches. In some cases, nanotech solutions appear to belong more to the realm of science fiction rather than science. For example, perhaps blood-borne robotic agents will someday be used to deliver treatment to infected cells.

The short-term implication of nanotechnology could well be driven by Moore's Law, which states that the minimum feature size shrinks by half every eighteen to twenty-four months; this would suggest that one could operate at a scale of 100 nanometers to 35 nanometers for the next ten years.

Recent advances in nano- and microfabrication suggest that today's tedious process of genome sequencing and expression profiling can be made dramatically more efficient by utilizing nanofabricated surfaces and devices. Expanding the ability to characterize an individual's genome will not only revolutionize the accuracy of diagnostics but also enable personalized medicine, that is, delivering the right drug for the right disease to the right patient. Research and development (R&D) will profit in a similar way through refined analytical tools that will allow studies from single molecule and cell manipulations up to studies of whole genomes and proteomes, complementing existing laboratory tools in the discovery of disease pathways and drug targets. However, beyond facilitating target discovery and diagnostics, nanotechnology can provide new formulations and routes for drug delivery, enormously broadening the therapeutic potential currently available. In general, these applications can be divided into three main categories: detection, diagnosis, and therapeutics. (See Table 3.1.)

DETECTION

Imaging

Many currently used clinical tests reveal the presence of a molecule or a disease-causing organism by detecting the binding of a specific antibody to the disease-related target. Traditionally, such tests are performed by conjugating the antibodies with inorganic/organic dyes and visualizing the signals within the samples through fluorescence microscopy or electronic microscopy. However, dyes often limit the specificity and practicality of the detection methods. Nanobiotechnology offers a solution by using semiconductor nanocrystals (also referred to as "quantum dots"). These minuscule probes can withstand significantly more cycles of excitations and light emissions than typical organic molecules, which more readily

Table 3.1
Specific Medical Applications for Nanobiotechnology

Detection and Sensors
• Sensor systems for the detection of pathogens (in vivo and ex vivo)

Clinical Diagnostics
• Rapid and inexpensive DNA characterization using nucleic acid arrays (the core competency of Nanosphere, Inc.)
• Development of new imaging technologies for earlier diagnostic of cancer and other diseases

Therapeutics
• Medical devices
  Effective and less expensive health care using remote and in vivo devices (e.g., monitoring of blood glucose levels and releasing of insulin correspondingly)
• Drug delivery
  New formulations and routes for drug (and gene) delivery
• Gene therapy
  Methods for correcting specific mutations in individual genomes using artificial proteins as "nanorobots"

---

decompose. This stability allows investigators to track ongoing, simultaneous cellular activities in tissue for longer intervals (due to the fact that various-sized nanocrystals can yield an entire spectrum of colors) (Drexler 2001).

Quantum Dot Corporation (Hayward, California) licenses technologies that load quantum dot particles that glow with sharp colors into single latex beads. These beads allow for an enormous number of distinct labels: each bead can attach to molecules of DNA composed of different sequences of genetic information and can be easily compared against a library of known sequences, creating a huge variety of distinct labels (nano "bar codes") for biological tests and diagnostic detection kits (Quantum Dot, Inc. 2004).

Individual Target Probes

Despite the advantage of magnetic detections, optical and colorimetric detections will continue to be chosen by the medical community. Nanosphere (Northbrook, Illinois) is one of the companies that developed techniques that allow doctors to optically detect the genetic compositions of biological specimens. Nano gold particles, studded with short segments of DNA, form the basis of the easy-to-read test for the presence of any given genetic sequence. If the sequence of interest is present in the samples, it binds to complementary DNA tentacles on multiple nanospheres and forms a dense web of visible gold balls. This technology allows the detection of pathogenic organisms and has shown promising results in the detection of anthrax, giving much higher sensitivity than tests that are currently being used (Nanosphere, Inc. 2004).

DIAGNOSTIC APPLICATIONS

Current diagnostic methods for most diseases depend on the manifestation of (visible) symptoms before medical professionals can recognize that the patient suffers from a specific illness. But by the time those symptoms have appeared, treatment may have a decreased chance of being effective. Therefore, the earlier a disease can be detected, the better the chance for a cure is. Optimally, diseases should be diagnosed and cured before symptoms even manifest themselves. Nucleic acid diagnostics will play a crucial role in that process, as they allow the detection of pathogens and diseased cells at such an early, symptomless stage of disease progression that effective treatment is more feasible. Current technology (such as polymerase chain reaction [PCR]) leads toward such tests and devices, but nanotechnology is expanding the options currently available, which will result in greater sensitivity and far better efficiency and economy.

Protein Chips

DNA microarrays are used to examine the changes in gene expression and to generate a database of patterns of gene expression or gene expression changes associated with a certain phenotype or physiological state. DNA chips, however, cannot be used to quantify protein expression levels. Proteins play the central role in establishing the biological phenotype of organisms in healthy and diseased states and are more indicative of functionality. Hence, proteomics is important in disease diagnostics and pharmaceutics where drugs can be developed to alter signaling pathways. Protein chips can be treated with chemical groups, or small modular protein components, that can specifically bind to proteins containing a certain structural or biochemical motif (Lee et al. 2002). This process becomes more specific and sensitive by using nanoscale fabrication techniques that allow the chemical components on the chip to self-assemble onto the proper regions of the chip. Two companies currently operating in this application space are Agilent, Inc., and NanoInk, Inc. Agilent uses a unique noncontact ink-jet technology to produce microarrays by printing oligos and whole cDNAs onto glass slides at the nanoscale. NanoInk, a Northwestern University start-up, uses dip pen nanolithography (DPN) technology to assemble structures on a nanoscale of measurement by literally drawing molecules onto a substrate (NanoInk 2004).

Sparse Cell Detection
Sparse cells are both rare and physiologically distinct from their surrounding cells in normal physiological conditions (e.g., cancer cells, lymphocytes, fetal cells, and HIV cells). They are significant in the detection and diagnosis of various genetic defects. However, it is a challenge to identify and subsequently isolate these sparse cells. Nanobiotechnology presents new opportunities for advancement in this area. Scientists developed nanosystems capable of effectively sorting sparse cells from blood and other tissues. This technology takes advantage of the unique properties of sparse cells manifested in differences in deformation, surface charges, and affinity for specific receptors and/or ligands. For example, by inserting electrodes into microchannels, cells can be precisely sorted based on surface charge. They can also be sorted by using biocompatible surfaces with precise nanopores. The Nano-biotechnology Center at Cornell University (NBTC) is currently using these technologies to develop powerful diagnostic tools for the isolation and diagnosis of various diseases (NBTC 2004).

THERAPEUTIC APPLICATIONS

In the year 2000, U.S. pharmacies processed 2.7 billion prescriptions for only two hundred of the most common drugs. During the same year, two hundred thousand Americans were hospitalized because of drug side effects. The most effective drugs act as "bullets," delivering the right dosage at the right time and at the right specific receptor; thus, they minimize drug toxicity. Nanotechnology can provide new formulations and routes for drug (and gene) delivery.

Drug Delivery

Nanoparticles as therapeutics can be delivered to targeted sites, including locations that cannot be easily reached by standard drugs. For instance, if a therapeutic can be chemically attached to a nanoparticle, it can then be guided to the site of the disease or infection by radio or magnetic signals. These nanodrugs can also be designed to "release" only at times when specific molecules are present or when external triggers (such as infrared heat) are provided. At the same time, harmful side effects from potent medications can be avoided by reducing the effective dosage needed to treat the patient. By encapsulating drugs in nanosized materials (such as organic dendrimers, hollow polymer capsules, and nanoshells), release can be controlled much more precisely than ever before. (Targesome, Inc., is using this technology to develop proprietary drugs.) Drugs are designed to carry a therapeutic payload (radiation, chemotherapy, or gene therapy) as well as for imaging applications (LaVan, Lynn, and Langer 2002).

Surfaces

In nature, there are a multitude of examples of the complicated interactions between molecules and surfaces. For example, the interactions between blood cells and the brain or between fungal pathogens and infection sites rely on complex interplays between cells and surface characteristics. Nanofabrication unravels the complexity of these interactions by modifying surface characteristics with nanoscale resolutions, which can lead to hybrid biological systems. This hybrid material can be used to screen drugs, as sensors, or as medical devices and implants. NanoSystems, owned by the Irish drug company Elan, developed a polymer coating capable of changing the surface of drugs that have poor water solubility (Elan Corporation, PLC 2004).

Biomolecular Engineering
Designing biomolecules is not a new concept, for chemicals and synthesized versions of proteins have been used for years. However, the expense and time involved in traditional approaches limit the availability of bioactive molecules. Nanoscale assembly and synthesis techniques provide an alternative to traditional methods. Improvements can be achieved due to the ability to carry out chemical and biological reactions on solid substrates, rather than through the traditional solution-based processes. The use of solid substrate usually means less waste and the ability to manipulate the biomolecule far more precisely. EngeneOS (Waltham, Massachusetts) pioneered the field of biomolecular engineering. The company developed the engineered genomic operating systems that create programmable biomolecular machines employing natural and artificial building blocks. These biomolecule machines have a broad range of commercial applications—as biosensors, in chemical synthesis and processing, as bioelectronic devices and materials, in nanotechnology, in functional genomics, and in drug discovery.

Biopharmaceuticals
Nanobiotechnology can develop drugs for diseases that conventional pharmaceuticals cannot target. The pharmaceutical industry traditionally focuses on developing drugs to treat a defined universe of about five hundred confirmed disease targets. But approximately 70 to 80 percent of the new candidates for drug development fail, and these failures are often discovered late in the development process, with the loss of millions of dollars in R&D investment. Nanoscale techniques for drug development will be a boon to small companies that cannot employ hundreds of organic chemists to synthesize and test thousands of compounds. And nanobiotechnology brings the ability to physically manipulate targets, molecules, and atoms on solid substrates by tethering them to biomembranes and controlling where and when chemical reactions take place, in a fast process that requires few materials (reagents and solutions). This advance will reduce drug discovery costs, will provide a large diversity of compounds, and will facilitate the development of highly specific

drugs. Potentia Pharmaceuticals (Louisville, Kentucky) is an early-stage company that is attempting to streamline the drug development process with the use of nanotechnologies (Harvard Business School 2001).

THE FUTURE OF NANOBIOTECHNOLOGY

Many nanotechnology applications are expected to come to the market around 2010 and will not be strictly related to biotechnology but will pave the way for the most exciting medical applications; for instance, enhanced materials, improved hard drives, optical networking, and computer chips will be the first applications for the new frontier in nanotechnology. But two types of medical applications are already emerging, both in clinical diagnosis and in R&D. Imaging applications, such as quantum dot technology (see the minicase on Quantum Dot Corporation), are already being licensed, and applications for monitoring cellular activities in tissue are coming soon. The second major type of application involves the development of highly specific and sensitive means of detecting nucleic acids and proteins. Companies such as Agilent, Affymetrix, Motorola, Nanogen, Illumina, Gene Logic, NEN Life Sciences, Perkin Elmer, and Invitrogen are active in this field (Milunovich and Roy 2001).

By 2010 to 2015, we will see that products being tested in academic and government laboratories will be creeping into commercialization. Clinical diagnostics and sensors will continue to advance. Sparse cell isolation and molecular filtration applications should, by then, make it to market. And some of the drug delivery systems will already be commercialized or in advanced clinical trials—for example, those being developed by NanoSystems (discussed in a minicase) or by American Pharmaceutical Partners, which is testing the encapsulation of Taxol, a cancer drug in a nanopolymer called paclitaxel.

Most medical devices and therapeutics are a decade or more away from market. Drug target manipulation as well as device implantations require a complex technical infrastructure, as well as complex regulatory management. EngeneOS, Inc. (the subject of a minicase), for instance, had difficulty in proving out its nanoantenna technology due to limitations encountered when transmitting radio-control data to the nanoparticles of interest (Hamad-Schifferli et al. 2002).

COMMERCIALIZATION ECOLOGY

The most important aspect of any technology is its eventual outcome. Will the product make it to the market? To answer this question, we need to look at the *idea agents,* who are the source of the innovation; the *transformation agents,* who can transform innovative ideas into reality; and the *conversion agents,* who ensure that the new technology is actually adopted and integrated into a marketable product. The fuel for ideas lies at the intersection of the federal funding agencies (the National Institutes of Health [NIH], the National Science Foundation [NSF], the Department of Energy [DOE], etc.) and the universities. The transformation agents are members of the venture capital and private equity community backing an entrepreneur. In nanobiotechnology, the academic technology transfer offices, which have learned from the experiences of molecular biology, have an opportunity to play an increasingly significant role. And finally, the conversion agents will eventually be responsible for turning ideas into profitable ventures. They are the ultimate decisionmakers who decide whether to embrace the new technology. Figure 3.3 shows how the various elements in the commercialization ecosystem interact (Branscomb, Kodama, and Florida 2003). Today, the most significant nanotechnology investments are made in the idea agent stage.

More than 90 percent of the biotechnology/life science companies have relationships with academic institutions, and 59 percent have sponsored research in the institutions. Agreements with universities are short-term arrangements and involve small amounts of money. The Quantum Dot Corporation case study illustrates the birth of a company that had its origin in university research and was commercialized by academic entrepreneurs.

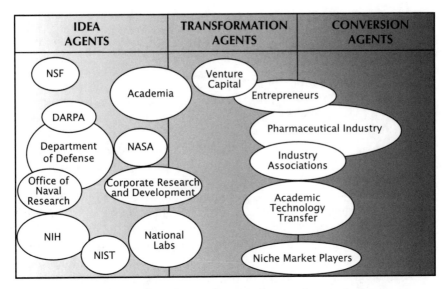

Figure 3.3   Schematic Depicting Different Elements Interacting to Form a Viable Nanobiotechnology Commercialization Ecosystem

Note: NSF=National Science Foundation; DARPA=Defense Advanced Research Projects Agency; NIH=National Institutes of Health; NISTA=National Institute of Standards Technology; NASA=National Aeronautics and Space Administration.

COMMERCIALIZATION STRATEGIES

The commercialization of nanobiotechnology is characterized by three distinguishing factors: (1) boundaries between research and commercialization are blurred, (2) transformation agents need to be highly technical, and (3) conversion agents need to devise multiple strategies.

There are several reasons why the research-commercialization boundaries are blurred. The explosion of information in fundamental knowledge has increased the areas in which nanotechnology can be applied (diagnostics, therapeutics, sensors, etc.). Some advances in these areas (such as the miniaturization of diagnostics) have an immediate market. However, others that have more potential are also riskier and are longer-term projects. Academic

Minicase: Quantum Dot Corporation

Quantum Dot Corporation was founded in 1998 in Hayward, California, by Dr. Joel Martin and Dr. Bala Manian. Martin was an accomplished entrepreneur and venture capitalist and formerly the president and chief executive officer of Argonaut Technologies, Inc. Manian, an entrepreneur and inventor, held over twenty-five patents and was the founder of Biometric Imaging, Lumisys, and Molecular Dynamics. Quantum Dot develops and sells "novel solutions to accelerate the discovery and development of functionally validated novel drug targets at the cellular level" (Quantum Dot, Inc. 2004).

The company's products and services utilize quantum dots, which are minute semiconductor crystals created at Lawrence Berkeley National Laboratory, the Massachusetts Institute of Technology, the University of Melbourne, and Indiana University. These nanoscale particles are characterized by unique properties that make them a superior detection platform for both nucleic acids and proteins. Quantum dots provide a tremendous variety of simple and inexpensive solutions for a wide range of problems in drug discovery and development. These problems include the need for more biologically relevant information with increased sensitivity, as well as easy-to-use, high-throughput, low-cost assays.

The company is currently developing high-sensitivity multiplexed assays that facilitate the measurement and explication of drug interactions both in vitro and at the level of cellular interactions. Toward achieving a goal of improving the drug discovery and development process, Quantum Dot Corporation is focusing on providing proprietary technologies through technology licensing and transfer programs to pharmaceutical and biotech partners. These solutions include Qdot™ nanocrystals custom-engineered for the requirements of specific biological assays. Key collaborations of interest include Quantum Dot's arrangements with Genentech, Inc., for the development of cellular imaging assays, and with GlaxoSmithKline, for genotyping assay development.

environments are usually more willing to accept such commercialization odds.

Entrepreneurs as transformation agents cannot be generalists and must be experts in the nanobiotechnology field. Technical evaluation requires an in-depth understanding of the technology (engineering, biotechnology, materials) and a real grasp of the market (medical needs). Today, few people or teams have the right expertise to thoroughly evaluate the technology.

More established firms as conversion agents manage the development of nanobiotechnology using multiple strategies. In certain cases, the existing technological assets impose a barrier to embracing new opportunities. This situation corresponds to quadrants III and IV in Figure 3.4. Depending on the extent to which resources must be invested to integrate the new opportunity into the existing process, the manager can choose to either facilitate creation of a new venture or continue to monitor the innovation with

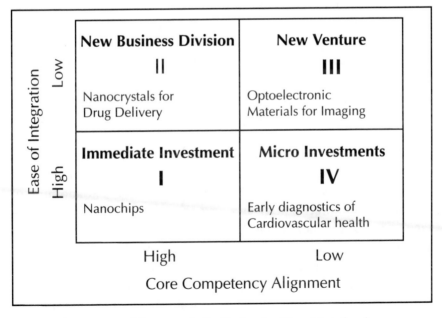

Figure 3.4   Managers' Framework for Evaluating Nanobiotechnology Opportunities

smaller targeted investments. For example, for a biodevice manufacturer, nanobiotechnology may not be an attractive proposition initially. In such a case, the firm could make minor investments in the technology to retain the option to exploit its uniqueness at a later time.

In other cases, a firm's technological assets might be complementary to a particular nanotechnology, and the firm will immediately embrace the opportunity. This situation corresponds to quadrants I and II in Figure 3.4. The task is more challenging if the short-term investments clearly outweigh the benefits. The firm must be aware that significant resources are often required to study the possibilities of the technology in the evaluation phase. Powerful resource coordination skills as well as a good understanding of the technology are key for commercialization success.

In addition, in the early development phase, companies need to take into consideration the diffusion strategies they will need to implement as the products move into their life cycles. A key starting point in creating a diffusion strategy is to identify the major segments of the technology adoption life cycle, as described by Gordon Moore (Moore 1999).

The major segments include the innovators, the early adopters, the early majority, the late majority, and the laggards. Although the actual representatives of these categories of the cycle may differ according to the specific market that the nanobiotech technology is addressing, there will be some commonalities across the markets.

The innovators will be the different university research labs, government research labs, and corporate labs experimenting with nanobiotech (see the Nanosphere, Inc., minicase). The early adopters will consist primarily of biotech firms that view adoption of the new nanobiotechnology as a way to build a competitive advantage. The early majority will be pharmaceuticals and large biotech companies that recognize the value of nanobiotechnology and adopt it to improve and maintain their leadership positions in the market. The late majority will be pharmaceuticals and medical device companies that are a little more skeptical of the new technology. Laggards will be the organizations that simply refuse to adopt the technology, possibly due to ethical issues with nanobiotech.

Minicase: Elan Corporation

Elan Corporation, PLC, was founded in 1969 in Dublin, Ireland, by Donald E. Panoz. The company began with a vision: "To approach the challenge of drug delivery from an entirely new angle—that of controlled absorption of a drug by the patient's body" (Elan Corporation, PLC 2004).

Today, Elan is a fully integrated biopharmaceutical company, with research activities focusing on Parkinson's disease, Alzheimer's disease, cancer, multiple sclerosis, pain management, and auto-immune diseases. The company concentrates its resources on the discovery, development, and marketing of products and services in neurology, pain management, oncology, infectious disease, and dermatology and on the development and commercialization of products using its extensive range of proprietary drug delivery technologies (www.elan.com).

To help develop its expertise in drug delivery, Elan acquired NanoSystems, LLC, a subsidiary of the Eastman Kodak Company, in 1998. Elan purchased from Kodak all the assets and liabilities of NanoSystems, LLC, for approximately $150 million (www.kodak.com). The pioneering work done by NanoSystems in the use of polymers to stabilize drug nanoparticles with poor water solubility for delivery to target sites was seen as a tremendous asset to Elan, helping to drive its innovative product pipeline.

Therefore, the strategy for the current nanobiotech firms should be to initially focus on university, government, and organization research labs. Beyond keeping the firm linked with all of the latest developments in the field, this strategy will put a given product in the hands of the gatekeepers. The labs will begin to "talk it up" at conferences, and the sponsoring companies will create a buzz around the technology. For their early customers, these firms must provide a real competitive advantage in using their products. Most of these firms' early profits will derive from differentiation and genuine innovation. At this stage of development, there is a high rate of concept failure, which companies must be prepared to handle. The case of EngeneOS clearly illustrates this point.

Minicase: Nanosphere, Inc.

Nanosphere, Inc., is a venture-backed and privately held life sciences company located just outside Chicago, Illinois. Chad Mirkin, Ph.D., and Robert Letsinger, Ph.D., two professors from Northwestern University, cofounded the company in 2000.

The proprietary technology at Nanosphere uses the unique properties of nanoparticles, and the company has developed a nanoparticle-based DNA detection system with "10 times more sensitivity and 100,000 times more specificity than current genomic detection systems" (Nanosphere, Inc. 2004). The technology is being positioned to complement and eventually eliminate the need for costly and time-comsuming PCR/fluorophore-based approaches, with the potential to dramatically simplify traditional analysis procedures. Nanoparticle probe detection systems will lower the cost, improve the quality, and markedly decrease the time to market for handheld molecular testing devices.

The company envisions itself becoming the "global standard" of molecular diagnostics. The overall estimated market potential of the three major applications of this technology—clinical research, in vitro diagnostics, and animal/food/environmental markets—is

At the same time, once a nanobiotech firm has developed an innovative, marketable technology, the company managers must focus their energies on forming a "dominant design" around their technology. They must ask themselves pertinent questions, such as whether the nanomaterial will be used first for a diagnostic or medical device, and their decisions may cause the firm to make some concessions. But once momentum builds, it will propel the dominant design firm up the technology S-curve and give it a substantial market leadership position.

Establishing a dominant design within the technology is essential to extracting large profits. Once an appropriate design emerges, a nanobiotech firm must focus on process, delivery, and service innovation. Meanwhile, technological differentiation among the rest of the nanobiotechs in a market will fade. The firm's skill sets will need approximately $7.5 billion. Thus far, Nanosphere has created a strategic partnership with IDEO Design and Development, headquartered in Palo Alto, California, to commercialize its first generation of detection systems.

Producing diagnostic devices could be a profitable business. The major obstacle the company's managers face is the need to convince researchers, both clinical and basic, that Nanosphere's technology is appreciably better than that which they are already using. Clinical researchers are especially hard to convince because diagnostic standards have been established using the current technology. To compound this problem, Nanosphere's managers will also have to explain to doctors how the data that are generated with the system correlate to the data the physicians are used to seeing and interpreting. Furthermore, companies that are already major players in the molecular diagnostic market, such as Roche and Applera, will most likely not adopt these new technologies to improve their products and maintain their position as market leaders. These problems can be overcome, but solving them will require extensive marketing research and the hiring of talented and knowledgeable medical directors to serve as liaisons to the medical community. There is always a reluctance to adopt new technology, but companies with vision and an understanding of early-entrance marketing can succeed.

to change once a dominant design is established as the company transitions from being science-driven to being market-driven.

Great technologies, though they almost always provide large returns for society, will not always produce large returns for the firms that developed them. This fact is an extremely important consideration for the nanobiotech industry, as it is currently made up of uncommercialized technologies. For a nanobiotech firm, no matter how revolutionary and important its technology is, the ability to reap returns from the technology will depend on a multiplicity of factors. The Wright brothers created the first airplane, but they reaped only a miniscule portion of the value derived from their innovation.

At the two ends of the spectrum, a commercialization strategy for a technology will either be competitive or cooperative. On one

Minicase: EngeneOS

EngeneOS was founded in late 2000 by NewcoGen Group, the Boston-based venture firm, with the hope of exploiting the opportunities resulting from the convergence of many disciplines (biology, physics, and computer science). EngeneOS was set up to develop technology and products to enable the design and building of programmable biomolecular devices consisting of hybrid natural/non-natural materials.

Back in January 2002 in the British journal *Nature*, EngeneOS reported the discovery of a "Nano antenna." Two months later, the company downgraded the project to pursue other potential applications of its technology platforms. What had happened?

Unfortunately, this action came as no surprise to scientists in the field, as technology often lags behind the vision of science. Nano antennae were created with the goal of controlling the conformation of biomolecules with radio-frequency energy, switching them from one state to another at will and thus combining the speed, efficiency, and cost of electronics with the specificity of biology. The problem encountered was that the energy generated by the antennae gold particles was simply not enough to change the conformation of the biomolecules; therefore, the hope of controlled delivery and activation of drugs using this method diminished. "Technologies fail all the time," said Carlo Rizzuto, Ph.D., a business development manager of EngeneOS, "however the idea is still viable and there are still potential applications that can come out of it" (Rizzuto 2001). Such problems are bound to plague companies in the early stages of nanotechnology. Eventually, the technology will catch up with the theories on which these companies are based. The pioneering work done by these firms will pave the way for the success of companies to follow, and if the early companies can survive, they will also reap the benefits of their current struggles.

Right now, EngeneOS has regrouped and is focusing on inlicensing technologies and building partnerships in the biochip area, hoping to take advantage of the vast bioinformatic information and its Engineered Genomic Operating Systems. These systems consist of component device modules supported by modeling and design tools. Initially, these development tools will probably be licensed to drug companies as a way of improving drug development. The long-term strategy of EngeneOS will be to actually develop nanoscale devices to realize the potential that seems like science fiction to most people today. In this context, it is worth noting that sequencing the human genome seemed like science fiction to most biologists, let alone laypeople, as recently as the early 1990s. The major question facing EngeneOS at present is whether the company can buy the time it needs to create the future it envisions.

hand, a competitive strategy involves direct entry into the product market, where the nanobiotech firm utilizes its technology to create a new value chain. On the other hand, a cooperative strategy involves working with established firms to integrate the unique value that nanobiotechnology brings to their own products. In determining the appropriate commercialization strategy, the nanobiotech firm should first assess the market in which it competes, using several criteria.

Appropriability (or the environment for exclusivity) is necessary for competitive protection. The imperative elements of appropriability for nanobiotech firms include intellectual property protection, secrecy (via trade secrets, noncompete clauses, and the complexity of the technology), and speed (in developing the innovation and in moving on to develop the next generation of that innovation). Maintaining an optimum appropriability environment in nanobiotechnology is difficult due to the speed of technology turnover.

The complementary asset environment is codependent on other firms' products and technologies. Complementary assets include competitive manufacturing, technologies, suppliers, services, and distributors. Such assets are held very tightly and are nearly impossible for a start-up firm to acquire. In particular, distribution channels, manufacturing, and established FDA regulatory knowledge and relationships are complementary assets.

The pharmaceutical market is one of the best existing case studies for a market of ideas. In this market, appropriability is high,

Minicase: NanoInk, Inc.

On April 4, 2002, NanoInk, Inc., a Chicago-based nanotechnology company, announced a $3 million first round of financing, led by the venture capital (VC) firms Galway Partners, LLC, and Lurie Investment Fund, LLC. Given the greatly diminished level of investment in the first quarter of 2002, what makes NanoInk particularly attractive to VC investors?

Founded in 2001 by Northwestern University professor Chad Mirkin, NanoInk is a nanotechnology company providing a platform process for nanoscale fabrication. "NanoInk will provide the tools and research collaborations that will lead to advances in diverse areas, from speeding drug discovery to building new types of electronics that are smaller and more powerful than those available today" (personal communication, March 2002).

NanoInk's patent-pending propriety technology, called Dip Pen Nanolithography((DPN), promises to be a valuable tool for companies and researchers entering the nanoscopic world, where products are smaller, faster, cheaper, and smarter. With the DPN technology, the company partners with corporations and research institutions in order to use proprietary nanofabrication technologies as manufacturing and research tools. In addition to its applications as a research tool, DPN will be a viable manufacturing tool that will enable the production of nanoscale structures by printing them on the desired platform, in much the same way a dot-matrix printer builds patterns on paper. Overall, NanoInk hopes to position DPN as an *industrial* process that is capable of building nanoscale devices with novel materials in a high-throughput manner.

NanoInk established itself primarily as a platform developer. This approach allowed the firm to build revenue in a relatively short period of time by selling tools and performing development services for larger companies. History has shown, however, that platform companies often do not have the long-term success that product-driven companies do, as it is often cheaper for larger companies to acquire such firms rather than to continue to pay for their services. There are two ways around this problem: developing a large customer base, thereby increasing revenues to the point where buyouts are infeasible, or developing products that can be sold directly to end users.

and the complementary assets are tightly held. This situation has allowed the same primary incumbents to lead the market for well over a hundred years, spanning many developments and technologies. These innovations ultimately only serve to reinforce the existing value proposition of the big pharmaceutical companies.

The optimal commercialization strategy for ideas in a nanobiotech firm in this market is to seek a cooperative contracting strategy. In doing so, the firm should mature its technology until its value within the market is obvious. The firm should then play various pharmaceuticals and mature biotech companies off against each other to create a bidding war for the ultimate sale of their company, which will maximize the returns of the nanobiotech innovation for the innovators. A summary of the inherent challenges and potential solutions is shown in Table 3.2.

Core competency identification and alignment are key considerations for the manager evaluating a nanotechnology idea, assuming it meets all other criteria (such as performance expectations, scalability, and cost-effectiveness). Based on the S-curve framework, we propose that issues of low customer acceptance and downstream product integration can be addressed by incorporating downstream product integration intelligence in product strategy, driving toward standardizations.

Several other challenges come with the commercialization of nanobiotechnology, including mobilizing academic entrepreneurs, handling regulatory issues, competing with the big pharmaceuticals, anticipating a competitive response from the incumbents in the industry, and funding issues. These challenges must be overcome before companies can become profitable or projects can have real applications.

Advances in enabling technologies have accelerated the onset and development of nanobiotechnology. As we have seen, compelling scientific and business drivers will accelerate the pace of innovation in this field and create significant commercialization opportunities. We believe that university and research labs are playing a decisive role in closing the gap between the research and commercialization frontiers for nanobiotechnology applications.

Table 3.2
Commercialization Challenges

Challenges	Explanation	Possible Solutions
Uncertainty of effectiveness of innovation	Considering the early stage at which most nanobiotech companies and innovations currently exist, the ultimate effectiveness of great ideas when taken from concept to product is unproven.	Avoid reliance on one key innovation for your company's success. This is especially true in the world of biotech and pharmaceuticals due to the large impact of regulatory agencies on the ultimate profitability of an innovation. Although a nanobiotech company may only have one great idea, it should develop a few additional ideas because failure does and will happen. Diversification is key to ensure long-run viability. Do not, however, lose focus on the primary innovation due to diversification, as this loss of focus will prevent a nanobiotech company from raising a second round of financing.
Scalability	Scalability is the ability to cost-effectively produce an innovative nanobiotech product and sell it efficiently to the mass market, or the product targets.	If the product will not scale, either (1) find the bottlenecks and engineer/innovate around them, (2) table grandiose plans for an initial public offering, (3) forgo venture funding opportunities and make plans to run a small company with a small niche, or (4) use sales of the unscalable niche product to fund the development of new products.
Funding	Funding is the ability to raise the capital necessary to pay for all the resources required to bring a concept to product and then a product to the market.	Since 2000, an enormous amount of money has been raised through venture capital and private equity funds. This money has generally not been spent, as it is being held back in anticipation of a rebound. Basically, for a nanobiotech company, this indicates that the money is available: it will just be a little more difficult to access than it was in 2000. Companies with great management teams and solid business models will have no problem raising funding. This will hold true across sectors. The first key to raising money is putting together a good management team. The team should ideally include proven managers with past success in running a profitable business and inventive, renowned scientists with past success in developing innovative biotech/Pharma technologies and/or products. The next key is putting together a coherent, solid business plan where the down-the-road profits are easily discernable even in the executive summary. This is currently a "hot" space, which works in the nanobiotech companies' favor. So raise more money than you need now, even at the expense of ownership, because what is hot ebbs and flows, and a nanobiotech company does not want to be caught in a situation where it needs money when the sector is in disfavor.

Table 3.2 (continued)

Challenges	Explanation	Possible Solutions
Getting innovation into big Pharma's pipeline	This challenge is defined as the ability to get meetings with decisionmakers in the large pharmaceutical companies so that products and technologies can be effectively pitched.	First, figure out who the decisionmakers are at the big Pharma companies. Title is not always indicative of key decisionmaking authority. Therefore, do due diligence on the company that involves more than just conducting research on the Internet. Simply talking to people in the company can lead to extremely valuable insights. After locating the decisionmakers, the key to getting meetings with them is to find allies either inside or outside the company who are trusted colleagues of the decision-makers, as cold calling rarely works. The network of fellow scientific researchers is the best place to start.
Scarce resources	This relates to the ability to acquire the scientific talent and components necessary to develop an innovative nanobiotech product.	First, locate within a biotech cluster (as there are not yet any nanobiotech clusters per se). This gives a nanobiotech company the choice of Boston, San Diego, and Palo Alto. To a lesser extent, Seattle and Research Triangle are also options. As discussed in the location strategy section, locating in a cluster provides many advantages. The biggest is the ability to attract and retain the scarcest resource in nanobiotech—Ph.D. nanotech researchers. Location is important in attracting talent for these reasons: most of the talented researchers already reside in these locations and are not particularly interested in moving to less desirable areas; by remaining in a cluster, innovative researchers can keep their options open in case the current nanobiotech company does not work out; and many of these individuals are tied to the universities located in these clusters.
Patience	The problem is not neces-sarily with nanobiotech companies. These companies are typically aware of the long and difficult road ahead to bring their products to market. Instead, this relates to not overselling the company to investors and ensuring that investors understand the cycle when entering the investment so their impatience does not derail a promising company.	Resist the urge to overhype or oversell a company or innovation to potential angel or venture capital investors. Although all entrepreneurs and scientists are understandably excited about their innovations and their potential impact on the field of health care, overselling this impact or the timing of this impact can create major problems down the road as investors become anxious to recoup their investments. This can lead to selling the company or innovation before the value of the innovation has even neared its peak.

REFERENCES

Branscomb, R., F. Kodama, and R. Florida. 2003. "The Impact of Academic Research on Industrial Performance." National Academy of Engineering Report, 77–114.

Drexler, E. 2000. "The New Nanofrontier." *Scientific American,* November 27. Available at http://www.sciam.com/article.cfm?articleID=000 C7091-590A-1C75-9B81809EC588EF21.

———. 2001. "Nanotech." *Scientific American* 285: 32–91.

Elan Corporation, PLC. 2004. Available at http://www.elan.com/ DrugDelivery/. Accessed June 2004.

Feynman, Richard. 1959. "There Is Plenty of Room at the Bottom." Presented at the annual meeting of the American Chemical Society, California Institute of Technology, December 29. Full lecture available at http://www.its.caltech.edu/feynman/plenty.html. Accessed on October 16, 2001.

Hamad-Schifferli, K., J. Schwartz, A. T. Santos, S. Zhang, and J. Jacobson. 2002. "Remote Electronic Control of DNA Hybridization through Inductive Coupling to an Attached Metal Nanocrystal Antenna." *Nature* 415: 152–155.

Harvard Business School. 2001. "Potentia Pharmaceuticals." June. Available at http://www.thebiotechclub.org/industry/businessplan/bplan2001.php.

LaVan, David A., David M. Lynn, and Robert Langer. 2002. "Timeline: Moving Smaller in Drug Discovery and Delivery." *Nature Reviews Drug Discovery* 1: 77–84.

Lee, K-B., S.-J. Park, C. Mirkin, J. Smith, and M. A. Mrksich. 2002. "Protein Nanoarrays Generated by Dip-Pen Nanolithography." *Science* 295: 1702–1705.

Milunovich, S., and J. Roy. 2001. "The Next Small Thing: An Introduction to Nanotechnology." Merrill Lynch Report, September 4.

Moore, G. 1999. *Inside the Tornado.* New York: HarperCollins.

NanoInk, Inc. 2004. Available at http://www.nanoink.net. Accessed June 2004.

Nanosphere, Inc. 2004. Available at http://www.nanosphere-inc.com. Accessed May 2004.

NBTC (Nano-biotechnology Center, Cornell University). 2004. Available at http://www.nbtc.cornell.edu/default.htm.

Quantum Dot, Inc. 2004. Available at http://www.qdots.com. Accessed April 2004.

Rizzuto, Carlo. 2001. Discussion with Kellogg Biotech Ventures class, March.

PART II

THE BUSINESS OF SCIENCE: ENHANCING THE VALUE OF THE INNOVATIONS THROUGH THE PERFECT BUSINESS MODEL

There are many opportunities for making money in biotechnology but only one way of creating a sustainable business—the old-fashioned way, by creating a unique value to society that can be captured as profits. Making biotechnology products is a long, expensive, and risky process. It takes twelve to fifteen years and costs more than $800 million per drug. Furthermore, only 15 percent of the drugs that enter preclinical trials make it to market, and barely 30 percent of the marketed drugs recover the costs of development. Rarely is the original "inventor" the one with the resources to bring the product to market. Most often, the invention will change hands many times before reaching the final customer, and with each transaction, there is a chance for somebody to make money. Considering that there are fifteen hundred biotechnology companies in the United States with each one pursuing different technologies or products, the number of potential transactions is immense. It is therefore not surprising that biotechnology has captured the attention of eager investors around the world.

Transactions can (but do not always) increase commercialization efficiencies, which is part of the reason why small biotechnology companies that lack resources opt to commercialize their products. By licensing their discoveries to seasoned pharmaceutical companies (which are expected to have the necessary commercialization knowledge), their products should reach the customers' hands faster and more economically. At the same time, both parties mitigate costs as well as the risks inherent in product discovery and development. Small biotechs of this type are known as platforms because they provide early-stage products, tools, or services to mature companies to be further developed into therapeutics or diagnostics. By selling services and tools, platform companies avoid the burdensome regulatory approval process and can reach profitability faster than integrated companies (companies that move products from discovery to market). In addition, they might receive royalties from the drugs that were developed using their tools or services, adding to their future profitability. However, as explained in Chapter 4, "Sustaining Platforms," these companies face important challenges, including the constant threat to commoditization, the complexity of dealing with multiple customers, and the need to achieve significant scale to remain competitive.

Giving away innovations to an experienced pharmaceutical company might not always be the best model on which to build a biotech firm. The perceived value of a biotech product increases as the risk of development decreases and as the product gets closer to launch (Figure II.A). Thus, by outlicensing inventions at the early stages of development, biotech firms relinquish much of their value to the pharmaceutical companies.

Moreover, many of the innovations require specialized processes that do not "fit" the mainstream capabilities of Pharma. This was the case in the early 1980s when pharmaceutical companies were too slow in incorporating specialized processes for the commercialization of therapeutic proteins. The field was then left open for the flagship biotech entrants, such as Genentech, Amgen, and Genzyme, to vertically integrate and develop commercialization capabilities for their protein-based drugs. These biotechs then provided value to the industry not only as product innovators but also as process innovators. A similar situation seems to exist today in the stem cell field, as companies involved in the discovery processes are finding themselves responsible for creating commercial scale as well. This situation is illustrated by ViaCell (Boston, Massachusetts), which is simultaneously engaged in the discovery of stem cell–based therapeutics and spearheading the scaling-up efforts.

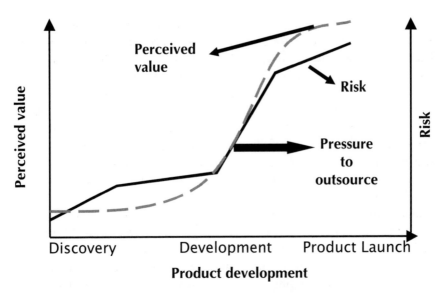

Figure II.A   Perceived Value and Risk of a Biotech Product Development

An alternative to giving the product away is to forward integrate—incorporating commercialization capabilities into the company's discovery competence. But this is not an easy task for a young, resource-poor biotech company. Diverting funding from research and discovery toward the building of a commercialization process not only distracts the company from its core research competency but also starves the research pipeline. Typically, the situation becomes a resource-balancing act, and the company is forced to bet on a single product and devote all available resources to developing it. The company then becomes a "one-horse show," with its survivability dependent on the success of that one product. Considering the odds in the drug discovery process, this strategy becomes a highly risky proposition.

Even when the product makes it to market, the company more often than not finds itself in a very vulnerable position. Unless the product is a blockbuster, the revenues will be insufficient to build a new pipeline. And with no prospects of new products coming out of the company pipeline, it becomes difficult to sustain growth rates. The small biotech firm then becomes a prime acquisition target—especially if the company's capabilities are not unique or innovative enough (unlike the Genentech, Amgen, and Genzyme situations) and can be easily duplicated and improved on by Pharma. Such was the case with COR Therapeutics (San Diego, California), which was acquired by Millennium Pharmaceuticals (Cambridge, Massachusetts) in 2001 for $2 billion just after COR launched its first product, Integilin, for acute coronary syndrome. Similarly, Scios, Inc. (Sunnyvale, California), was acquired by Johnson and Johnson (New Brunswick, New Jersey) in 2003 for $2.4 billion after launching its first product, Natrecor, for congestive heart failure. Both firms developed a successful product but not sustainable companies. Clearly, regardless of whether or not these biotechs were sustainable, investors in both transactions were very successful.

If forward integration is difficult, backward integration might be impossible. Still, biotechs continue to pursue this strategy. Instead of beginning by building the company around a technical innovation, many start-ups focus on developing the commercialization capabilities first, with the hope that by inlicensing innovations or acquiring companies, they will eventually build a deep and continuous pipeline. These companies are generally known as non-research-development-only (NRDO) start-ups. The question is, What is the *unique* value that NRDOs provide to the industry and society? One could argue that a start-up can never have the manufacturing scale and marketing reach that Big Pharma has, unless the products involved include specialized processes or small niches that do not fit Pharma's capabilities. If that is the case, then the barriers of entry for performing development are low. But if the barriers are low for NRDOs, they should also be low for the inventors of the innovation, so why should inventors outlicense to NRDOs?

The poster child for the NRDO model is Pharmion (Boulder, Colorado), which was founded in 1999 and concentrates on licensing, developing, and commercializing therapeutic products in hematology and oncology. Pharmion acquired its most visible product, thalidomide, from Celgene (Warren, New Jersey), buying the rights to commercialize the drug in Europe and Asia. Thalidomide was first introduced in the 1950s and was prescribed for nausea and insomnia in pregnant women. However, it was found to cause severe birth defects in children whose mothers had taken the drug in their first trimester of pregnancy. After that, no pharmaceutical company had any interest in exploring alternate indications or even being associated with thalidomide, given its history. Celgene, founded in 1986, then had an open opportunity to forward integrate to commercialize thalidomide in the United States (the biggest market). The drug, although very risky, had the potential to be an effective cancer treatment. But Celgene had no global capabilities and sold Pharmion the rights to develop the drug in the rest of the world. Pharmion was then left with arguably the less-attractive market and is now marketing the drug (Pharmion Thalidomide) in Australia, New Zealand, and Turkey.

There is not a clear formula to build a sustainable biotech firm (Table II.A). Outsourcing models work in some cases; vertical integration works in others. An alternative common approach for biotech firms is to use a mixed model, wherein companies do both outsourcing and integrating either in sequence or in parallel. The strategy then becomes a highly sophisticated and company-specific one, in which decisions need to be choreographed regarding when and how to outlicense and integrate.

Among the companies pursuing a mixed model are Abgenix (Fremont, California) and Sangamo Bioscience (Richmond, California). Abgenix, founded in 1996, is applying the sequential mixed model and is involved in fully developing human therapeutic antibodies. Abgenix has over fifty academic, biotechnology, and pharmaceutical companies as partners. Although it initially focused on preclinical and phase I research, Abgenix is currently expanding its efforts into phase II, phase III, and the manufacturing stages. Clearly, it is moving forward to integration.

By contrast, Sangamo is searching for sustainability through a parallel mixed model. Founded in 1996, Sangamo licenses its zinc finger transcription factors (known as ZFT) technology for the screening of small molecules, monoclonal antibodies, and protein drug leads to other biotechnology and pharmaceutical companies. ZFT are transcription

Table II.A

Major Advantages and Challenges of the Platform, Integrated, and Mixed Models

Model	Advantages	Challenges
Platform	• Development risk mitigated • Achieves faster profitability than integrated model	• Short technology shifts • Capturing respectable profit margins • Managing multiple customers • Achieving scale
Integrated	• Captures the lion's share of the product's value • High profit margins	• Highly risky (expensive) • Takes a long time to profitability
Mixed	• Combines advantages of both platform and integrated models	• Creates significant organizational complexity and sophistication

factors that can be used to activate or repress gene expression. Sangamo uses the revenues generated through the licenses to advance its own therapeutics.

Regardless of which model a company uses to enhance its value—platform, integrated, or mixed—sustainability can only be accomplished if the company is offering a *unique* value. Value is created with a revolutionary product (reinventing the value chain or creating new markets) or with an evolutionary product or process (making products or processes faster or better). If the company is developing a revolutionary product, it will enhance its value by competing on the added benefits of the product; if the company is improving the products or processes, it will need to enhance value by competing through cost and scale. If the product offers no real value, then the company becomes a short-term transaction (Figure II.B). This transaction could be very profitable, but it will be vulnerable to the markets. Many of the biotechs today fall into this third category.

Sustainability is challenging even in the presence of real value creation. In Chapter 4, we look at the unique challenges of platform companies and provide some guidance on how they can best be sustained, even in difficult markets. In Chapter 5, "Mergers and Acquisitions as a Strategic Alternative in Biotech," we consider whether acquisitions could be a valid strategy to extend sustainability.

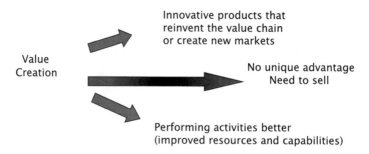

Figure II.B   Value Creation

Note: There are two ways of creating value in the biotechnology sector, either by creating innovative products or by performing activities better than others. The value of innovative products is enhanced by highlighting the benefits of those products. The value of improved processes is enhanced by creating scale and competing through cost. If the company is not adding any unique value to the industry, it has no competitive advantage and will not be sustainable.

Chapter 4

SUSTAINING PLATFORMS

Erich Hoefer, Sriram Jambunathan,
Anne Tse, and Mariona Vincens

Platform companies are firms that provide enabling technology or services to companies that are developing pharmacological therapeutics. These firms typically focus on the drug discovery process and generate revenues through licensing fees, grants, and research and development (R&D) collaboration fees. The companies enjoyed a high degree of investor interest in the late 1990s, enhanced by the excitement of the completion of the Human Genome Project.

However, a major shift has taken place in recent years, and platform companies now find themselves under extraordinary pressure from investors to shift their strategies to developing therapeutic products instead of discovery tools and services. This trend was already evident in 2002 when Invitrogen acquired InforMax for less than 80 percent of its cash value. Hyseq followed suit and acquired Variagenics for 87 percent of its liquid cash positions, showing that investors were actually expecting these companies to put their cash and short-term investments in negative net present value (NPV) or value-destroying projects.

Regardless of their profitability, platform companies play a significant role in maximizing the efficiencies of the commercialization process in the biotech and life science industries, providing specialized and innovative tools to the drug development sector. Platform company revenues, however, have not grown as fast as experts had initially projected. It is not clear whether the failure of these companies is based on a fundamental flaw in the platform model.

In this chapter, we analyze the main hurdles that platform companies face and offer some solutions to achieving long-term sustainability. Our approach is based on numerous interviews with executives and investors in the biotechnology industry. Our analysis points to four major barriers that platform companies must surmount to ensure their long-term sustainability: industry structure, scalability, structural risk, and financial pressure. As an example, we will use the company Illumina from San Diego, California. Finally, we will offer some strategies to augment the sustainability of platform companies.

INDUSTRY STRUCTURE LIMITS PROFITABILITY

The platform market is fragmented and competitive. The number of platform companies has grown exponentially since the late 1990s, with hundreds of companies offering tools and services that range from bioinformatics to toxicogenomics to interference RNA, and so on. One reason for this explosion of platforms is that industry entry barriers for platform companies are relatively low due to several factors, including the following:

- The initial investment to start up a platform company generally comes in the form of grants from the National Institutes of Health (NIH), Small Business Innovation Research grants (SBIRs), and grants from other government and university entities.
- No platform technologies—whether in tools or research process—have yet been able to attain a dominant position, thereby leaving room for new entrants to compete.
- Time to market is relatively short because platform technologies do not have to go through lengthy regulatory approvals. Incumbents therefore do not enjoy any structural first-mover advantages. As a result, companies offering scientific talent with promising technologies are attracted to the industry.

Although multiple platforms are being offered to improve the drug discovery process, the number of very large buyers is limited. Almost all platform companies aim to license their technology or to provide services to big pharmaceutical companies and a handful of larger biotech product companies. Most of them are competing to gain the attention of a Pfizer or Novartis. Moreover, as pharmaceutical companies continue to consolidate, platform companies will only face fewer and bigger customers who can exercise bargaining power over pricing or simply purchase the platform company outright. The latter is becoming more common as pharmaceutical companies consider the growing problem of royalty stacking—the need to have access to multiple platforms, each associated with a royalty price tag. If a pharmaceutical firm is to pay a 2.5 percent royalty for the use of five different technologies, the big company will find itself in the position of having to spend 12.5 percent of the final product's sales when it reaches the market (Clark 1999).

Platform companies face harsh competition from other companies, but their biggest competitors are government entities and universities that can offer the same tools at a fraction of the cost. For example, universities today are a rich source of monoclonal antibodies. In 2004 alone, the National Institutes of Health granted over three thousand monoclonal antibody projects and more than eight thousand deoxyribonucleic acid (DNA) chip projects to universities and other research institutes (CRISP Database 2004). This overwhelming amount of resources and a lack of distinctiveness reduces buyers' switching costs and thereby increases competition.

SCALABILITY IS CRITICAL TO SUCCESS BUT DIFFICULT TO ACHIEVE

Scale is key for the success of platform companies, for several reasons. First, given the industry's structure that limits high margins, volume becomes the only route to profitability. Second, scale provides cost advantages as platform companies spread their R&D investments and other fixed costs over larger volumes and/or additional product/service categories. This effect is particularly significant in firms with high operating leverage, which is the case for most platform companies. Third, scale improves the ability of firms to compete by creating a natural barrier to entry and ultimately a positive effect on profitability. Finally, scale enhances firms' ability to raise additional capital.

However, platform companies need to surmount three important hurdles in order to reach scale: (1) achieving widespread market acceptance, (2) building in technological flexibility to accommodate new applications and avoid being leapfrogged, and (3) ramping up production capacity rapidly in a cost-effective manner.

Achieving Widespread Market Acceptance

Widespread acceptance of a technology occurs when there are both technological and market validations. Technological validation determines the extent to which customers believe in the effectiveness of the technology relative to products currently on the market. Market validation determines the commercialization potential for the technology or how much of the value the firm can capture with its technology.

Technological validation is achieved through academic collaborations that signal the viability of the platform to the scientific community. Having a technology platform that is used and validated by academic peers aids in the acceptance of the technology by industry. In this context, a venture capitalist we interviewed described the importance of having hard data behind a technology application, rather than relying on theory alone.

One firm that is pursuing this strategy is Gene Logic, Inc. (Gaithersburg, Maryland). Gene Logic incorporated in 1994 and commenced commercial operations through collaborative research efforts in 1996. The company launched its primary bioinformatics product, the GeneExpress System, in December 1999, providing both tools and services. By 2004, Gene Logic had almost 100 customers and 420 projects from government, universities, and biotechnology and pharmaceutical companies. About 50 percent of all pharmaceutical companies had partnered with Gene Logic, offering a validation for its tools.

Market validation is obtained though deals and collaborations with pharmaceutical companies that signal the commercial potential of the products developed using the platform. A platform company can capture a percentage of downstream revenues either by providing a key enabling technology to allow discoveries to be made or by improving R&D productivity by reducing time and cost in the drug discovery process. Market validation from the big players in the industry positively affects the valuation of platform companies and leads to greater acceptance and scalability of the technology. For instance, when Isis Pharmaceuticals announced its collaboration with Lilly on its "antisense" platform in 2001, share price increased 50 percent in the same day and 70 percent in less than a week (Yahoo Finance 2003).

BUILDING FLEXIBILITY

Scalability is achieved when diversity is pursued in the applications for which a particular platform can be used. By extending a technology across multiple applications, the cost of developing that technology can be spread across multiple products. In the case of IGEN International (Gaithersburg, Maryland), recently acquired by Roche, its electrochemiluminescense technology is applicable across a range of industries, from life sciences to industrial and diagnostic markets. Hence, it can be used as a key component in high-throughput screening in the drug discovery process and also in *E. coli* pathogen diagnostic tests in meat products.

One way of building in flexibility is by providing a bundle of solutions on the platform. For example, the GeneExpress System of Gene Logic comprises a broad range of products and services based on data derived from gene expression studies of human and animal tissues and cells, and it can be used to understand gene expression patterns, genetic abnormalities, disease pathways, disease mechanisms of action, and mechanisms of toxicity. Another example is provided by Genoptix (San Diego, California). This company built a biomedical platform that utilizes moving waves of light to investigate the biological properties of living cells and thereby develop specific diagnostics tests that are complementary to the drugs being discovered to treat a disease. If the platform can be used to develop both the diagnostics and the drug targets, its applicability and acceptance will increase.

Another way that flexibility provides an advantage to a platform company is by protecting it from being leapfrogged by other technologies. To sustain a low-margin strategy, a platform company must be able to respond to advances in competitive technology with a minimum of new investment. If the firm is too heavily invested in a product with a very narrow application, it may find it has to abandon its earlier technology investments and start over. But if it constantly builds flexibility into its products, the platform company can respond quickly to competitive threats.

Incorporating flexibility into the company's products must be a long-term commitment. Platform companies need to have a continuous influx of follow-on product/solution offerings in order to accommodate new applications and preempt competition. One reason why many promising platform companies falter is because their managers have not thought through the means to keep the technology useful and alive in new applications.

RAMPING UP PRODUCTION CAPACITY IN A COST-EFFECTIVE MANNER

The third hurdle to scalability is the ability to ramp up production capacity, for both products and services, rapidly and efficiently. To move a technological application from the lab to full-scale production, firms are required to commit investment capital and build operational capabilities beyond basic research.

At the same time, there is a trade-off between automation and flexibility. The more a platform invests in automation, the more inflexible the firm becomes. As critical as automation is to achieving scalability, investments in automation also make the firm more vulnerable to being leapfrogged by competing platforms. As a result, large investments in automation may quickly become an outdated liability for the firm. This observation applies equally well to the technology the firm invests in to achieve automation capabilities. Early on, most platform firms find that their enabling technologies have quickly become obsolete. For these reasons, investment timing is critical.

MANAGING STRUCTURAL RISK IS KEY TO CAPTURING VALUE

Drug discovery remains a risky and lengthy process, with the costs involved in developing a drug from discovery through the approval process averaging more than $800 million; moreover, the process lasts approximately fifteen years (Boston Consulting Group 2001). Approximately $520 million and 6.1 years are spent on drug discovery tasks such as identifying and validating potential disease targets and developing and optimizing potential therapeutic candidates in preparation for clinical trials. A critical driver of the development cost is the high level of risk involved in this process: as much as 75 percent of the cost is due to drug attrition, with only around five out of five thousand early-stage compounds making it to preclinical trials.

PARADOX: THOUGH POSITIONED TO CREATE VALUE, PLATFORM COMPANIES ARE UNABLE TO CAPITALIZE ON THE OPPORTUNITY

Just as all inefficiencies create room for value creation, a lengthy, risky, and costly drug development process presents a tremendous opportunity for any player who can enhance and expedite the process. Platform companies are well positioned to create and capture this value. However, they generally capture only a small portion of the total value created. The question, then, is why platform companies are unable to capitalize on this potential. We will explore four key reasons in the following sections.

PLATFORM TECHNOLOGIES DO NOT OFFER A CLEAR WAY
TO LOWER THE RISK OF THE DRUG DEVELOPMENT PROCESS

One major reason platform companies are unable to capitalize on potential is risk. Being at such an early stage in the pharmaceutical process and remote from the final product, platform products incorporate an enormous level of unavoidable risks, which must be passed down the value chain. The larger the risk to be transferred, the less value the platform company can capture from a pharmaceutical. Their customer must take into account the overall risk profile of the drug discovery process—not just the particular stage/process that the platform company addresses.

As an example, assume a platform company was able to offer best-in-class gene expression data. Any potential partner interested in developing a therapeutic based on those data would still have to invest in R&D to identify and validate a potential target, identify potential leads using high-throughput screening techniques, optimize promising leads for effectiveness, and test those leads in preparation for clinical trials. The partner firm has a high chance of losing its investment at later stages in the discovery process.

Even if the platform were able to minimize the risk of its product through the discovery phase, a significant amount of investment will still need to be made to develop a drug candidate through clinical trials. The combination of these risk factors reduces the amount of value that a pharmaceutical partner is willing to give up. Even a substantial reduction of risk at an early stage of development corresponds to only a small portion of a large pharmaceutical company's total financial risks. Therefore, even a well-validated platform technology may not be able to capture a significant part of the value if it remains focused on only one segment of the value chain.

PLATFORMS FACE COMPETITION FROM ALL QUARTERS

In addition to the competitive pressures discussed previously, competition makes it very difficult for platform companies to capture the value they are creating from mitigating risk. On the one hand, pharmaceutical companies are constrained in the total amount they can economically invest in drug discovery. At the same time, uncertainty in the efficacy of platform technologies drives pharmaceuticals to hedge their investments by partnering with multiple platform companies. Fundamentally, this means a given platform company must compete with platforms at all other stages of the discovery process for a scarce source of funding.

THE COMPANIES' FINANCING NEEDS AND THE HIGH DEMANDS OF FINANCIAL MARKETS MAKE LONG-TERM SUSTAINABILTY HARD TO ACHIEVE

Because of the lack of profitability and the large investment needed to implement their business models, platform companies rely heavily on equity markets to fund their operations. As a result, these companies face pressure from either private investors or public markets to deliver the expected returns on investment if they want to ensure their funding and sustainability. Therefore:

- Most private companies become public when the initial public offering (IPO) window of opportunity is opened, with the objectives of meeting their financing needs and responding to the pressure of venture capital investors to cash out their investment. These companies may take this avenue even if they are not positioned for long-term sustainability at the time.
- Public companies, though still far from achieving profitability, feel pressure to meet the high expectations of the financial markets in order to continue funding their operations through second offerings.
- Valuations of platform companies are more vulnerable to the capital markets than those of product (drug) development companies. Although valuations of product companies correlate with clinical success, valuations of platform companies are a reflection of the biotech index (Figures 4.1 and 4.2).

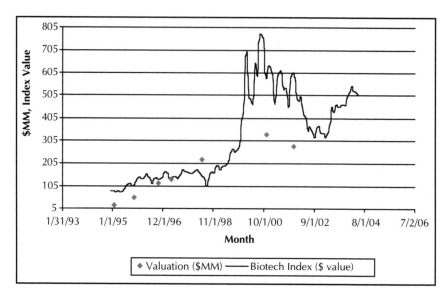

Figure 4.1   ArQule's Valuation versus the Biotech Index, 1995–2004

Note: ArQule is a fee-for-service life science company with therapeutic programs in kinases and ion channels. Valuations reflect the biotech index. $MM = million dollars.

Source: Research by Justin Busarakamwong, Mike Joo, and Paul Seamon, Center for Biotechnology, Kellogg Graduate School of Management.

A LOWER VALUATION IS AWARDED TO PLATFORM TECHNOLOGY COMPANIES AS OPPOSED TO THOSE WITH WELL-DEFINED PRODUCT (DRUG) DEVELOPMENT

Even if it is difficult to identify outperforming biotech companies when looking at financial figures such as revenues and net income, some of the older and larger companies have been performing extraordinarily well in the financial markets.

The drug discovery industry is demanding technology platforms and content services that can deliver value by helping to turn data into knowledge that will then translate into viable novel therapeutic and diagnostic solutions. The drug gains value with each step, and though approved drugs have substantial worth, data, tools, and services are not worth as much. Figure 4.3 shows the sig-

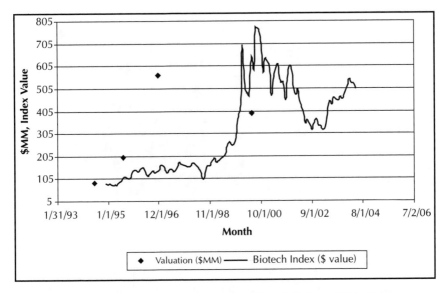

Figure 4.2   Neurogen Valuation versus the Biotech Index, 1994–2004

Note: Neurogen is a product and drug development company focused on therapies for inflammation, depression, and obesity. Valuations increase when clinical milestones are met. $MM = million dollars.

Source: Research by Justin Busarakamwong, Mike Joo, and Paul Seamon, Center for Biotechnology, Kellogg Graduate School of Management.

nificantly lower value (more than two times lower) of tool-based companies relative to their cash. Similarly, Figure 4.4 illustrates the ratio of market capitalization to annualized revenue for platform-based companies in comparison with the biotech index. Clearly, platform companies are valued significantly lower (more than two times lower) relative to their cash and three times lower than the biotech index. As a result, long-term sustainability is hard to achieve.

Since 2000, financial markets have increased their pressure over biotech companies, and a fundamental change has occurred in the way the biotech industry is valued: biotech companies have begun to be judged on how they deliver on their promises about revenues and earnings.

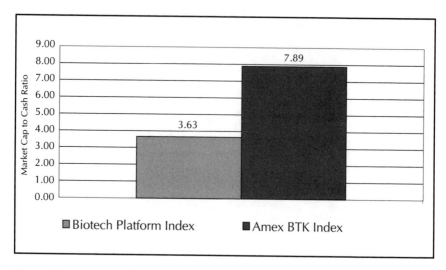

Figure 4.3   Average Ratio of Market Capitalization versus Short-Term Investments of the Seven Largest Platform Companies Compared with the Amex Biotech Index

Source: Mike Blastick and Sheldon Ng. Data obtained from Yahoo Finance, Kellogg Center for Biotechnology, Kellogg Graduate School of Management.

Note: The figure shows the significantly lower value (more than two times) of platform-based companies relative to their cash, compared to the companies listed with the Amex Biotech Index (BTK).

Because of the financial needs and the pressure of private investors, most companies become public when they are still far from reaching profitability. Most of them raise enough cash in the IPO to fund their operations for eighteen to twenty-four months, but it is uncertain whether they will be able to reach profitability before additional funds are needed. If these companies are forced to return to the public markets to raise additional funds, they may find that investors will only respond to a significant shift in strategy toward the higher returns (and risks) of therapeutic development.

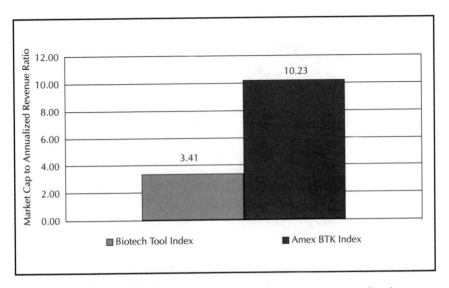

Figure 4.4 Average Ratio of Market Capitalization versus Annualized Revenue of the Seven Largest Platform Companies Compared with the Amex Biotech Index

Source: Mike Blastick and Sheldon Ng. Data obtained from Yahoo Finance, Kellogg Center for Biotechnology, Kellogg Graduate School of Management.

Note: The figure illustrates how platform companies are valued significantly lower (three times) than biotech companies referenced by the Amex Biotech Index (BTK).

WHAT CAN PLATFORMS DO?

To achieve long-term sustainability, platform companies must overcome a number of obstacles. The industry in which they compete is highly fragmented, suffers from low barriers to entry, and faces strong buyer power in the form of relatively few large customers. To compete in this environment, it seems clear that platform companies must adopt a high-volume/low-margin strategy based on a distinctive technology advantage. And to attain the scale required for this strategy, they must first gain market and technology validation, pursue flexibility in product development, and invest in automation.

Minicase: Illumina—Succeeding through Automation

Illumina (San Diego, California) is a leading developer of next-generation tools for the large-scale analysis of genetic variation and function. The company's tools provide information that can be used to improve drugs and therapies, customize diagnoses and treatment, and cure disease. Illumina's products primarily impact the drug discovery process right after the target validation stage. Its technology platform also extends to other applications, such as high-throughput screening of pharmaceutical compounds and chemical detection. The main customers are pharmaceutical companies that develop drugs.

Understanding genetic variation and function is critical to the development of personalized medicine—a key goal of genomics. And multiple companies with competing technologies are targeting overlapping subsegments of the genomics market.

One of Illumina's primary components is its BeadArray™ technology—an optical fiber–based microarray test platform that offers benefits in terms of higher throughput and scalability. In addition, Illumina has the Oligator™—an oligonucleotide manufacturing capability that complements its BeadArray™ technology. The company leverages these technologies in the services arm of the business.

Differentiation through Distinctive Value Proposition

The genomics tools industry in which Illumina is a player is very fragmented and encompasses multiple competitors, each of which has competing technologies. There are few end users with applications and scale that justify the use of the relatively expensive platforms. Typically, big pharmaceutical companies and fully integrated biotechnology firms such as Amgen are potential buyers, and they wield a good deal of power in commercial relationships.

The barriers to entry for newcomers who aim to leapfrog existing technologies are generally low. Illumina may be able to compete successfully in the genotyping and gene expression markets, which total over $1 billion and are growing, because the company's leading proprietary technology decreases the cost for screening single nucleotide polymorphism (SNP) by 40 percent, dropping it from 50¢ to 30¢ per test. (Cost is one of the current barriers to screening all possible SNPs.) With its technology, Illumina intends to further reduce this cost to less than 10¢ per test. At that level, it would be financially feasible for big pharmaceutical companies to include this screening stage in the drug discovery phase to reduce attrition and thereby significantly lower the cost of drug development. This situation would also create entry barriers for new competitors.

Gaining Value by Scaling Up Production

Illumina's BeadArray™ technology is highly scalable and can be used to make a significant impact on the drug discovery process right after the target validation stage. By arranging the arrays in a pattern that matches the wells of industry-standard microtiter plates, Illumina can simultaneously process up to 3 million assays, throughput significantly beyond the capability of most of the technology currently available in the market (Illumina 2003).

To gain widespread acceptance in the genotyping market, Illumina needs to make its platform the industry standard for low cost–high volume gene expression testing. It has a partnership with GlaxoSmithKline Beecham to codevelop and market applications using Illumina's platform. Currently, its alliance with Applied BioSystems leverages its sales force and distribution channels to place and service instruments at client sites. However, the company has not yet made huge inroads in achieving broad industry acceptance.

Illumina leverages its technology by developing new products. A number of follow-on applications are being contemplated to ensure a diversity of applications for its technology platform. These applications will help Illumina augment its business and gain benefits of increased scale. One application is in the diagnostics market for tests where the number of patients being tested is large; this application will benefit from the company's automation capabilities. The key challenge will be to locate tests on the platform that meet the higher sensitivity requirements of diagnostics while leveraging the higher throughput capabilities of the BeadArray™ platform.

Furthermore, Illumina is adding capacity and investing in greater operational capabilities. The company has made investments of $0.5 million to create additional manufacturing capacity and $1.2 million to improve operational capabilities to support the BeadArray™ and Oligator™ technologies. These investments need to pay off in the form of increased revenues from the oligo synthesis services business and also from product sales. Currently, revenue as a proportion of R&D expense is very low and needs to be ramped up.

(continues)

Managing Structural Risk Is Key to the Ability to Capture Value

As mentioned, Illumina's technology makes its impact after the target validation stage in the drug discovery process. Genotyping improves the success rates for Pharma drug development by increasing drug specificity toward certain segments of the population. Illumina's technology has the potential to capture value by targeting this inefficiency in the discovery process. As a result, certain drugs in the clinical process that would fail when targeted at the larger general population could be "kept alive." Assuming the cost per test drops below 10¢ an SNP, integration of this technology into the drug discovery process could reduce the downstream risk of drug development for pharmaceutical companies in a cost-effective manner, allowing Illumina to capture a portion of the Pharma company downstream revenue.

The impact of Illumina's technology in enhancing R&D productivity has not yet been validated, and it is unclear what proportion of the end product value pharmaceutical companies will be willing to share. As a platform company that only affects one part of the value chain for Pharma companies, Illumina's impact may be marginal at best in reducing the overall risk of drug development; hence, it can lay claim to only a small or negligible proportion of the downstream revenue.

Illumina Must Resist Pressures from Investors to Abandon Its Platform Strategy Prematurely

Illumina was founded in 1998 with a seed round of $750,000. It was initially formed by Larry Bock of CW Group, and $8.5 million was raised from CW Group, Tredegar Investments, Venrock Associates, and ARCH Venture Partners. In January 2000, Illumina completed a $28 million private equity financing.[1] In July 2000, the company went public, raising $96 million.

At the time of the IPO, the company's main revenue streams came from government grants, and the capability to manufacture and commercialize its products was not fully implemented. As of December 30, 2001, the accumulated deficit was $50.1 million, and total stockholders' equity was $106.8 million. These losses principally occurred as a result of the substantial resources required for the research, development, and manufacturing scale-up effort required to commercialize its products.

If we look at the current expenditures of the company, we observe that R&D expenses account for over 60 percent of the total costs. Illumina recognizes that the research and development expenses, including facilities-related costs, are expected to increase substantially in future years to support research programs, internal product research, and technology development. In addition to this investment in R&D, Illumina is also expecting to incur substantial and increased selling, general, and administrative costs as the company begins to build up the sales and marketing infrastructure to expand and support product sales. Looking at the future, Illumina's task will be to manage the financial expectations.

Outlook for Illumina

One of Illumina's key strengths lies in its ability to offer a differentiated product, which will allow it to compete effectively despite the challenging industry structure. Furthermore, Illumina has made significant progress in building scalability by leveraging its technology across multiple products and pursuing market validation for those products. In terms of the company's future prospects, we would offer the following comments:

- Illumina must continue to reduce the cost of its technology platform, such that the cost per test in the gene expression market drops below 10¢ per SNP. This will help the company capture and become the dominant technology platform in the personalized medicine space.
- The evolution of the nascent gene association market and the value pharmaceutical companies place on the information generated through Illumina's technology will be a key determinant of the company's growth.
- Illumina will have to manage financial resources to work on multiple projects at the same time and ensure a sustainable pipeline of follow-on product offerings.
- The firm's ability to continue raising capital, especially in light of its reduced stock valuation, will be a critical factor.

Even when firms are able to build the needed scale, they must still overcome barriers related to the risk inherent in drug discovery, which significantly reduces their ability to capture fair value for their products. Finally, depending on its financing history and current needs, a platform company may find that its access to financing is restricted unless it abandons its platform strategy to pursue a product development strategy.

Given these significant hurdles, there is little wonder that so many platform companies have changed strategy to pursue therapeutics. Yet, based on our analysis of Illumina and other biotechnology firms, we believe that long-term sustainability is indeed possible for platform companies. The following sections outline a number of strategies that such companies must pursue.

HORIZONTAL CONSOLIDATION

In a fragmented market, there are always opportunities for creating value through consolidation, and for platform companies, consolidation brings several benefits. First, it addresses challenges associated with the current industry structure. By consolidating, companies reduce industry fragmentation and may therefore be better positioned to capture value from their customers. Consolidation also creates barriers to entry by increasing the minimum size necessary to compete.

Second, consolidation improves efficiency. As mentioned in earlier sections, platform companies have high fixed costs. Consolidation may allow two companies to increase their customer base, assuming that the new firm is effective at cross-selling its technology across its new customer base.

Mergers and acquisitions (M&A) activities have already kickstarted growth in the product arena, beginning with MedImmune's bid for Aviron and Cephalon's acquisition of Group Lafon. These deals were then quickly followed by Millennium's acquisition of COR Therapeutics, which was later overshadowed by Amgen's acquisition of Immunex. The wave of consolidation will spread to platform companies as they realize that given the sizable disparity between capacity needs and available capacity for platform solutions, consolidation is the most efficient way to achieve scale.

TECHNICAL AND MARKET VALIDATION

A coordinated strategy to validate the platform's technologies should be pursued through alliances with academics and market leaders alike. Such validation can have a significant, positive impact on a firm's ability to scale up and may also help to improve its ability to compete with other available technologies.

LEVERAGING EXISTING TECHNOLOGY ACROSS MULTIPLE PRODUCT AREAS

Another way to achieve scale, particularly over the short term, is to focus on developing add-on products to augment the platform company's existing technology. Because it is essentially pursuing a low-margin strategy, a sustainable firm will be able to maximize revenues relative to R&D expenditures. Building such flexibility into the platform's product development may also protect against leapfrogging.

INVESTING IN AUTOMATION WHEN APPROPRIATE

When pursuing a high-volume/low-margin strategy, investment in automation capabilities is critical. Only through the reduction of operating costs will a platform company be able to generate long-term returns. At the same time, investments in automation may put the firm at greater risk of being leapfrogged. As a result, the firm should pay particular attention to timing issues before committing to an investment.

VERTICAL CONSOLIDATION

One way to mitigate the effects of risk on the platform's ability to capture value is to consolidate multiple platform technologies around a core capability. By bringing multiple stages of drug discovery into a comprehensive toolkit, a platform company can add value by removing a greater portion of risk. An example of a firm that has taken this approach is Millennium Pharmaceuticals. Starting from a base of its own proprietary technologies, the company has invested in numerous technology alliances in order to offer a comprehensive drug discovery product to its customers. According to Millennium, as much as 95 percent of the technology that it offers to its partners comes from inlicensing agreements it has negotiated with other platform companies. In this way, it is able to capture more of the value in the discovery process by offering a comprehensive technology product to its partners and customers.

MANAGING FINANCING NEEDS TO MINIMIZE PRESSURE FROM INVESTORS

Finally, to be sustainable, a platform company must manage the expectations of its investors. Venture capitalists and stockholders alike generally expect very high returns from biotech firms, but given the issues addressed earlier, platform companies will be hard-pressed to meet those expectations. As a result, they should be wary of going to these financing sources. Instead, a sustainable strategy may involve pursuing alternative funding sources, such as collaborations with large pharmaceutical companies.

NOTES

1. New investors included the Tisch Family Fund, Lombard Odier and Cie, State Farm Automobile Insurance Company, Chase Capital Partners, PE Corporation, Dow Chemical Company, and Chevron Technology Ventures.

REFERENCES

Boston Consulting Group. 2001. "A Revolution in R&D: How Genomics and Genetics Are Transforming the Biopharmaceutical Industry." Boston Consulting Group Report, November, p. 6.

Clark, Andrew. 1999. "Inside the Fund Manager's Mind." *Nature Biotechnology* (supplement) 17: 29–30.

CRISP Database. 2004. National Institutes of Health. Available at http://crisp.cit.nih.gov/crisp/crisp_query.generate_screen. Accessed June 2004.

Illumina. 2003. www.illumina.com. Accessed December 2003.

Yahoo Finance. 2003. Available at http://biz/yahoo.com/ic/biotrx.html. Accessed December 2003.

Chapter 5

MERGERS AND ACQUISITIONS AS A STRATEGIC ALTERNATIVE IN BIOTECH

*Andrew Ingley, Gregory Lief, and Jeanne Lukacek**

Historically, funding patterns for the global biotechnology industry have been highly volatile and cyclical. Every few years, there are periods of ample capital availability through venture funding and public capital markets (primarily through initial public offerings [IPOs]), followed by periods with low company valuations, no exit scenarios for early-stage investors, and little capital market interest in biotech companies. At any given time, many biotech companies are nearing the end of cash reserves generated during the last "open" funding environment and are depleting cash reserves at a rate at which they will be unable to survive another six to twelve months. Based on our financial analysis of 200 public biotech companies, we found a full 35 percent will not survive the next twelve months.

Because of the vagaries of funding, biotechnology companies often shut down or rapidly change business models to generate life-sustaining revenues. Currently, there are over 400 public biotechnology companies worldwide and significantly more private companies. For a variety of reasons, biotech chief executive officers

Research overseen by Chris Ehrlich at InterWest Partners, Menlo Park, California.

(CEOs) are generally gun-shy about mergers and acquisitions (M&As) as a tactic for survival. They are usually reluctant to admit their companies are failing, and there is the fear of relinquishing the CEO position to a counterpart during a merger.

Given the economic funding environment for public biotech companies, we hypothesize that a strategic M&A is a viable and attractive alternative for at-risk companies in order to prolong a company's life and improve shareholder value. In this chapter, we will evaluate this hypothesis and recommend general guidelines for companies that wish to engage in strategic M&A. We will begin by analyzing the survivability of a representative selection of public biotech companies and then assess current funding sources for biotech companies. Next, we will evaluate a selection of recent mergers and acquisitions in the biotech industry, by way of case studies, in order to identify success factors in biotech M&A. We will conclude with general guidelines for companies currently facing a funding crunch and considering a merger or acquisition.

ANALYSIS

Like most young industries, the biotech sector is volatile. Exuberant periods are typified by high valuations and an abundance of funding, but they are followed by dismal periods when companies are undervalued and funding is scarce. Between 2001 and 2002, the biotechnology industry, as well as just about every other industry, seemed to be undergoing a period of depression. Sharp decreases in IPO and secondary offering volume, adverse money flows in mutual funds, and anemic venture investments were the major factors contributing to this backdrop, and biotechnology companies were hard-pressed to raise operating capital from sources other than internally generated cash flow.

The first section of our financial analysis examines the situation of public biotech companies in 2002 with respect to available case and burn rates. A significant proportion of public biotech companies had little cash remaining during that time. We selected 2002 because it was the most recent slump the industry had lived through; this same analysis could be made for 1992 or any other period when the capital markets were closed. The following sections of this financial analysis look at recent and historical trends in the capital markets that were causing a cash crunch for biotechs in 2002. We then examine the specific avenues of funding in light of recent trends related to biotechnology and other relevant industry investing. We feel the bulk of this analysis covers three areas of capital infusion that present the biggest challenges to struggling biotechnology companies: mutual fund money flows, public offerings, and venture capital. This analysis illustrates the difficulties public biotech companies face in raising funds despite their need for cash.

Our primary intent in conducting this financial analysis was to understand the financial status of publicly funded biotechnology companies during 2002. We expected to find a significant number of these companies with very little cash, relative to burn rates, due to the poor capital markets funding situation. Without significant cash infusions or alterations in operating and alliance strategies, these companies had very little time to perform before running out of cash. We then identified a "survivability index" that approximated the amount of time, in years, remaining for a company based on burn rate and current cash and short-term investments.[1]

By 2002, given their third quarter burn rates, 16 percent of the companies analyzed had less than six months of cash remaining, and a full 35 percent would not be alive in twelve months. A breakdown of the number of companies in each survivability category is presented in Figure 5.1.

By the end of 2002, all major sources of capital (mutual funds, public offering, and venture) for the biotechnology sector were diminishing. Discontinuous fund outflows can have severe and far-reaching implications. A net reduction in the amount of investable money that mutual fund managers can direct leads to three major consequences. First, the public markets suffer due to a lack of buying support and potentially some selling pressure as fund managers liquidate part of their holdings to cash out any redeeming shareholders. Second, the venture capital community suffers as fund managers allocate fewer dollars to venture capital

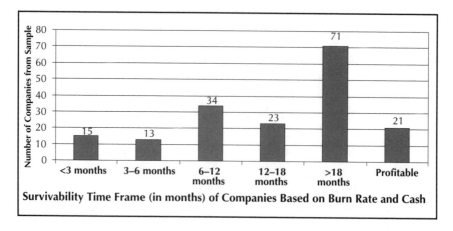

Figure 5.1   Biotech Company Survivability (sample of 177 total companies, third quarter, 2002 financials)

limited partnerships. And third, a lack of capital among institutional investors further depresses the IPO and secondary offering markets, which can further dampen an already tepid venture capital environment as public offering exits become less of an option. Overall negative signals in the mutual fund environment have a magnified impact on biotechs that may need capital in the foreseeable future.

If the mutual fund environment was considered bad for biotechs in 2002, then the public offering markets were a disaster. Data gathered from the Equidesk database detailed the number of priced IPOs and secondary offerings over the preceding twelve years. The data in Figure 5.2 illustrate that in 2002, the industry was experiencing a "closed" period in which biotechnology companies were hard-pressed to convince public market investors of the value of the firms.

The immediate consequence of a closed public market is that biotechs have one less avenue to pursue when seeking capital. A longer-term consequence is that venture investors are less willing to invest in companies when the exit timing is less clear. Because IPOs represent just such a strategy, the closed public markets will also serve to dampen sources of venture capital.

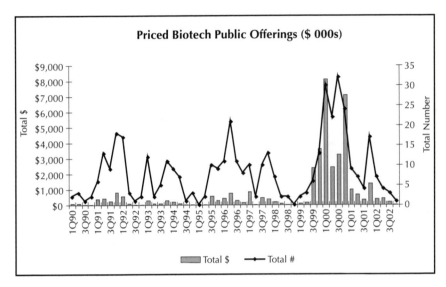

Figure 5.2   Historical Biotechnology Public Offerings, 1990–2002

Source: Figure drawn from data obtained with Equidesk database.

Note: 1Q = first quarter; 3Q = third quarter.

Withdrawn deals are also a good indicator of the health of capital markets. During 2002, the number of withdrawn deals grew, roughly equaling the number of priced deals (Figure 5.3). This suggests that even if management and investment bankers believed a company had good prospects for attracting capital, markets were not as positive.

Historically, venture capital has been the most accessible source of funding for biotechnology companies. A decline in capital spending by corporations, a lack of public offering activity, and a lukewarm mergers and acquisitions environment are commonly key drivers for a slowdown in venture capital investing.

The funding to start up biotech companies in the boom of 2000 dramatically slowed down in the subsequent years (Figure 5.4). By 2002, most of the reduced activity in funding was focused on late-stage investments and medical devices. Of 647

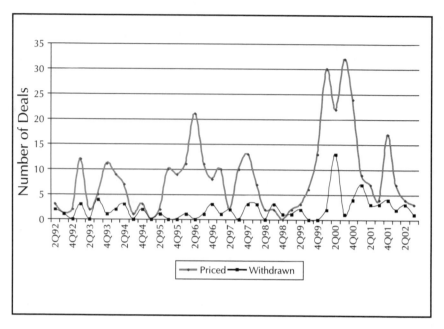

Figure 5.3   Public Biotech Offerings, Priced versus Withdrawn, 1992–2002

Source: Figure drawn from data obtained with Equidesk database.

Note: 2Q = second quarter; 4Q = fourth quarter.

companies that received financing in the third quarter of 2002, only 159 received first-time financing, the lowest number since the fourth quarter of 1994.

However, in 2002, the scientific fundamentals for biotech were stronger than they had ever been up to that point in history, and more products were in the pipeline than ever before. But with the cost to develop a single drug reaching $800 million, the major sources of capital dwindled and the cash reserves of biotech firms were drawn down. Most companies were facing relatively short survival windows. Given this bleak outlook, myriad companies needed to find alternative survival mechanisms.

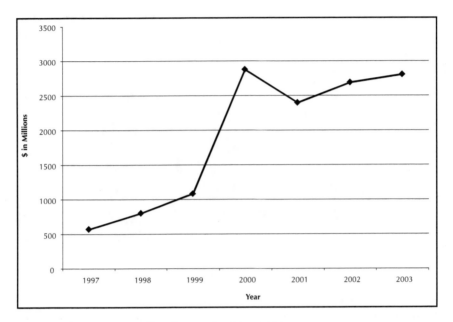

Figure 5.4   Venture Dollars Invested, 1997–2002

Source: Graph was drawn from data obtained from Steven G. Burrill (Burrill 2003).

MERGERS AND ACQUISITIONS AS A CAPITAL-
BUILDING ALTERNATIVE

In the period between 1996 and 2002, the health care industry experienced a rise in the number of M&A transactions and in the size of the average transaction, with a potential trend reversal occurring in a twelve-month period when all other sources of capital diminished (Figure 5.5).

We hypothesized that M&A activity might be a viable and attractive alternative for at-risk companies in terms of prolonging company life and improving shareholder value. To support this hypothesis and help managers and advisers understand what criteria

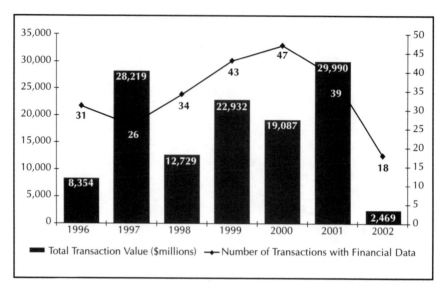

Figure 5.5   Biotechnology M&A Activity for Twelve-Month Periods from 1996 to 2002 Ending June 30

Source: Data from Thompson Financial DSC Platinum Database of transactions involving publicly traded biotechnology, health care equipment, or pharmaceutical companies. Data represent completed transactions where the acquiring company ended up with at least 50 percent of the outstanding shares postacquisition.

might improve the probability of a successful merger, we turned to historical M&A examples to glean key insights about why the transactions were undertaken and what factors led to their success.

We identified three measures to determine whether a merger was a successful alternative for biotech companies (targets):

1. Market value: Did the value of the combined company increase relative to the Amex Biotech Index?
2. Retention of key projects: Were a significant percentage of the key projects of the target company continued in the postmerger entity?
3. Retention of target management: Was a significant percentage of the key management of the target company retained in the postmerger entity?

The success of a merger or an acquisition cannot be judged by improved shareholder value alone, for project-based and organizational needs must also be met. Target management and employees have expended time and effort in their commitment to the projects within their portfolio, and therefore, they have a vested interest in seeing those projects continue. If the target company's projects are all discontinued after a merger, the company will essentially disappear. In such a case, one could argue that the merger has been a failure.

As for management retention, one reason mergers fail to come to fruition or fail after they have been completed is due to the problem of competing CEOs. If there is a place for the target management team in the new entity (or if the target management team wants to exit from the new entity), the probability of a successful merger increases. This was the situation in the merger of Arris Pharmaceuticals and Sequana Therapeutics. Kevin Kinsella, the CEO of Sequana, was a well-known and highly successful seed-stage venture capitalist. When John Walker, Arris's CEO, met with Kinsella, he perceived that Kinsella was not interested in being CEO of a biotech firm over the long run. Thus, Walker thought that one M&A hurdle—competing CEOs—would not be an issue in the merger (Longman 1997). There are, of course, some exceptions to the rule about competing CEOs. For example, if a top manager is looking for a way out of the organization, his or her retention in the new company will be less significant than continuing marketed and pipeline projects and boosting shareholder return.

To better understand the practical issues in M&A transactions, we turned to historical M&A cases to glean insights about why transactions were undertaken and what factors led to their success (Table 5.1). Six transactions were examined, in light of three "success" measures: the financial returns realized by target shareholders, the retention of target management after the merger, and the retention of key target projects after the merger. Our intention was to draw some basic guidelines that managers could consider prior to engaging in a transaction. These companies were selected because they gave us a sample that was diverse in

Table 5.1
Highlights of Merger Deals of Arris Pharmaceuticals—Sequana; Gilead
Sciences—NeXstar Pharmaceuticals; MedImmune—US Bioscience;
Cephalon—Anesta; Alliance Pharmaceuticals—Molecular Biosystems; Celera
Genomics—Axys Pharmaceuticals

Strategic Rationale	Acquirer	Target	Closing Date	Deal Value
Downstream integration	Arris Pharmaceuticals	Sequana Therapeutics	January 1998	$166 M
Cost avoidance, pipeline expansion, and international sales	Gilead Sciences	NeXstar Pharmaceuticals	July 1999	$600 M
Entry into a new therapeutic area	MedImmune Cephalon	US Bioscience Anesta	November 1999 October 2000	$580 M $444 M
Gain leadership position in market	Alliance Pharmaceuticals	Molecular Biosystems	December 2000	$11.6 M
Change in business model	Celera Genomics	Axys Pharmaceuticals	November 2001	$99 M

terms of the strategic reasons for the M&A activity. We focused on the strategic motivations behind the M&A and looked for critical success factors that could be applied to other companies considering such activity. Clearly, a survey of only six companies is inherently limited, but these firms represent good illustrations of the general strategic rationale behind biotech mergers. A summary of the major outcomes of the six analyzed cases is shown in Table 5.2.

ATTRIBUTES OF THE DEAL THAT MAXIMIZE SUCCESS

The following sections discuss a number of "attributes of the deal" that target companies need to evaluate when considering an M&A transaction:

Table 5.2
Case Studies Synopsis

Merger	Did the Merger Increase the Target's Shareholder Value?	Did the Target Company Have Key Management Retained?	Did the Target Company Have Key Projects Continued?	Was the Merger Successful?
Arris/Sequana	No	No	No	No
Gilead/NeXstar	Yes	Yes	Yes	Yes
MedImmune/US Bioscience	Neutral	No	Yes	Yes
Cephalon/Anesta	Yes	No	Yes	Yes
Alliance/Molecular Biosystems	No	Yes	Yes	Somewhat
Celera/Axys	Neutral	No	Yes	Neutral

Source: For each merger, we created a case study based on research from SEC filings, analyst reports, general press, and *In Vivo*. The goal of the case studies was to capture standardized information on the company's basic financials, business model, product/therapeutic focus, R&D activities, prior M&A experience, and management team chanages. In addition, we analyzed target company stock performance from thirty days prior to the announcement of the merger through the closing date of the transaction.

SALES FORCE

If the target firm has a sales force, especially one that is active in a given region, it should be acquired by a company with a sales force gap or a near-term product launch in that region.

Benefits

In this situation, there is a greater probability of retaining management and personnel postmerger and a higher likelihood of sustained financial success after the merger due to complementarities.

Example

Gilead's acquisition of NeXstar was done in part to leverage NeXstar's European sales operations for Gilead's product. Judged by the measurements we used in our analysis, this merger was a success.

Minicase: Arris Pharmaceuticals and Sequana Therapeutics

In late 1997, the trend in the biotech industry was for drug discovery companies to become vertically integrated—to have an in-house technology to identify novel targets and a way to generate drugs for those targets. Arris's CEO, John Walker, believed a company that achieved this goal created the most value for shareholders and for potential strategic partners. Prior to pursuing Sequana, Arris investigated joint venture relationships with other companies focused on generating novel targets but decided the structures of these deals were too strategically restrictive (Longman 1997).

Meanwhile, the proprietary technology owned by Sequana was facing competition and potential commoditization. According to Scott Salka, the former chief financial officer of Sequana, his company was still years away from introducing a product. Unable to convince investors that its technology was unique, the company agreed to be acquired by Arris (Longman 1991).

This transaction created a new company called Axys that had in-house capabilities extending from genomics to drug development. The transition was a simple vertical integration merger. Arris had no source of novel targets, but through the acquisition, it gained technology that would feed its drug targets. And for its part, Sequana gained resources to develop therapeutics for the targets identified through its genomic capabilities.

Overall, this merger should be considered a failure. Although it made logical sense to both entities, the market reacted negatively to the combined company that resulted, and Axys lost nearly half its value the month after the merger was announced. In addition, only one of eight key personnel remained with Axys after the merger. Two of the five corporate partnerships that were the primary sources of revenue for Sequana remained active one year postmerger. Two years later, Axys sold Sequana's pharmacological and agricultural capabilities. And four years after Sequana was acquired for $166 million, Axys was bought by Celera Genomics, a subsidiary of Applera Corporation, for $99 million—60 percent of the price paid for Sequana in the merger.

Minicase: Gilead Sciences and NeXstar

NeXstar was a fully integrated biotech company with two products on the market: AmBisome for severe fungal infections and DaunoXome for Kaposi's sarcoma. NeXstar's lipid encapsulation technology was instrumental in improving the delivery and ameliorating the side effects of drugs that had been on the market for years.

Gilead was also a fully integrated biotech company, focused primarily on HIV, hepatitis B, and influenza. We speculate that the Gilead management team was interested in NeXstar both for its portfolio of marketed products and for its lipid encapsulation technology. The antifungal and chemotherapeutic focus of NeXstar would complement Gilead's therapeutic focus, and NeXstar would provide Gilead with a European sales operation that it could use to leverage pipeline products as they came to market. Also, as BancBoston Robertson Stephens estimated, the acquisition was expected to drive Gilead to profitability nearly a year earlier than anticipated without the acquisition (King and Tenthoff 1999; Copithorne and Somaiya 1999). Achieving profitability sooner could drive additional shareholders to invest in Gilead, thereby creating additional value for the company.

Overall, the merger of Gilead and NeXstar can be considered successful. NeXstar shareholders received significant premiums on stock relative to the Amex Biotech Index through the beginning of 2000, at which point returns were on par with the index. Further, Gilead retained the majority of NeXstar's products, and a decent number of NeXstar's senior managers became vice presidents at Gilead.

THERAPEUTIC AREA

The target firm should have a product in development in the same therapeutic area as the buyer.

Benefits

In this scenario, there is a high likelihood of products being retained in the postmerger entity. With the right buyer, the

Minicase: MedImmune and US Bioscience

US Bioscience (USB) was a company that inlicensed late-stage oncology compounds, and it had internal manufacturing and marketing capabilities. Over the two years prior to the merger with MedImmune, USB had entered into discussions with several other potential merger partners, which made US Bioscience a willing target for the right acquirer (MedImmune 1999).

MedImmune was a fully integrated biotech company with three marketed products and five more in clinical trials. The firm focused on four therapeutic areas—infectious diseases, transplantation medicine, autoimmune disorders, and cancer (MedImmune 1998).

We speculate that the most important factor considered by the USB board in pursuing the merger with MedImmune was that management had wanted to sell the company for two years and that none of the prior merger discussions had resulted in a significant transaction. MedImmune offered to keep most of the USB operations intact, which, in the end, cost USB a premium, as reflected by its market value (MedImmune 1999). The merger greatly benefited MedImmune in its attempt to enter the oncology market. US Bioscience would instantly add a full oncology marketing infrastructure and two approved oncology products, thereby making MedImmune a true player in the marketplace.

Overall, the merger of MedImmune and USB can be considered a success in that the major products of USB were continued and even expanded after acquisition by MedImmune, despite the mediocre stock performance that ensued.

combined company's financial success should also be greater than that of the sum of its parts.

Example 1

The Cephalon/Anesta merger was unambiguously successful in our view, and the primary reason was that the target company, Anesta,

Minicase: Cephalon and Anesta

Anesta was a fully integrated biotech company with one internally developed product on the market—Actiq, an oncology product marketed through a contract service organization collaboration and manufactured through an agreement with Abbott. In addition, Anesta had outlicensed its former rights to the U.S. market. Anesta also had a drug delivery platform (OTS) that was being used to deliver pain management therapy to cancer patients. As such, Anesta was building out an oncology sales and marketing infrastructure.

Cephalon was also a fully integrated biotech company, and its premerger products were designed to treat epilepsy, migraine pain, Parkinson's disease, and excessive daytime sleepiness associated with narcolepsy. The company was interested in expanding into the oncology therapeutic area and at the time had a pipeline of three oncology products but no experience or infrastructure to support oncology product sales.

Cephalon wanted to leverage its resources in marketing Actiq. By acquiring Anesta, Cephalon was able to gain immediate access to a sales infrastructure in the oncology market, and it acquired the ability to hire experienced oncology sales personnel, specifically experienced in selling Actiq. After the merger, Actiq sales increased by over 250 percent, and Cephalon retained a legitimate presence in the oncology market.

Overall, the merger of Cephalon and Anesta can be considered successful. Although no key managers continued at Cephalon in significant roles after the merger, the sales of Actiq were very successful. The marketplace responded positively, and Cephalon shareholder returns were significantly greater than those of the Amex Biotech Index.

had a marketed product and infrastructure that fit perfectly with Cephalon's desire to become a player in the oncology market. The acquirer had stated its interest, had pipeline products in oncology, and wanted to leverage Anesta's established market force. The marketplace rewarded the shareholders for this strategic move.

Minicase: Alliance Pharmaceuticals and Molecular Biosystems

Before merging with Alliance Pharmaceuticals, Molecular Biosystems, Inc. (MBI) was in fairly poor shape. The company was down to three employees, had discontinued all R&D efforts, and had been through a failed merger attempt in 1999. MBI did have one product on the market—an intravenous ultrasound agent, Optison, that it had outlicensed to Mallinckrodt, Inc. (It eventually transferred complete control of the product to Mallinckrodt.) MBI needed to take some sort of action, whether it was a liquidation of assets or a merger.

Alliance's business model was focused on developing a product in-house and then finding a partner to take the product to the market. The company had been working on a product similar to Optison, but it lagged behind MBI's product by several years. By acquiring MBI, Alliance acquired a leadership position in a relatively crowded space. In the following months, Alliance continued the collaboration with Mallinckrodt in an amended form.

Overall, this acquisition should be considered fairly successful. Although the financial analysis indicates that Alliance stock subsequently underperformed the Amex Biotech Index, this was mainly due to factors endogenous to Alliance, not the MBI acquisition. The Optison product was continued postmerger, and one of MBI's three employees continued at Alliance as vice president of operations.

Example 2

The Alliance/Molecular Biosystems (MBI) merger was a similar situation. MBI's Optison product allowed Alliance to advance much more rapidly within its stated therapeutic area. Without the merger, Alliance would not have been able to launch its internal compound for a number of years.

CASH IS KING

The target firm should be offered cash or compensation through other nonstock transactions.

Minicase: Celera Genomics and Axys Pharmaceuticals

After the acquisition of Sequana in 1998, Axys continued its business model of discovering and developing drugs in-house and partnering with Big Pharma for clinical trials and for bringing the drugs to market. It had three large Big Pharma collaborations before the Celera merger, but all products were in the preclinical stage. Due to the overall market conditions in 2001, Axys stock was trading significantly below its 1998 level.

Celera was a traditional genomics tool company that was known for sequencing the human genome. In 2001, there was a significant trend (and pressure) for tool companies to become product companies, similar to Vertex and Millennium. With the purchase of Axys, Celera abruptly changed its business model to become a drug and diagnostics company. Axys helped to provide the biology and chemistry capabilities needed for Celera to diversify upstream.

Overall, the merger of Celera and Axys can be considered neutral. No key Axys management personnel stayed on at Celera in significant roles, but Axys projects and collaborations were continued. The marketplace initially responded positively to the merger, then turned sour on it. One year after the merger, Axys shares were performing on a par with the Amex Biotech Index.

Benefits

Cash is received up front, and the target firm does not have to worry about the subsequent stock performance of the buyer.

Example

The relative strength and future performance of a buyer is often difficult to judge. In the case of Alliance and Molecular Biosystems, had MBI been able to predict the future performance of Alliance, it would have been better off with a nonstock deal. We were not able to discern whether MBI considered a nonstock deal in the merger with Alliance; given its own precarious financial condition, it is possible that MBI did *not* weigh this factor at the time of acquisition.

WHO IS BEHIND THE DEAL?

It is important to have a clear understanding of the motivations and stability of the buyer's board and management team.

Benefits

Executive management turnover can be extensive when a company changes business models, which often negatively affects stock performance. If approached by a buyer likely to significantly alter its business model in the near term, the target firm *may* see enhanced stock returns (assuming a stock transaction) with a company whose board and management team see eye to eye on the changes required for transitioning to the new model.

Example

In the Celera/Axys merger, there was a significant change in management at Celera only two months after the merger was completed. In January 2002, Celera announced the resignation of J. Craig Venter as its president, and in April 2002, it announced that Kathy Ordonez, who was also president of Celera Diagnostics, would replace him. Also in January 2002, Celera announced the appointment of David Block as the chief operating officer of its therapeutics business. And in July 2002, Celera announced the appointment of Robert Booth as its senior vice president of research and development to lead its therapeutics R&D efforts (Applera Corporation 2002; Longman 2002). Management change contributed to the poor performance of Celera stock during the following six months.

THE MODEL

The buyer should be looking to forward integrate and be wary of backward integration.

Benefits

Forward integration generally (though not always) shows progress up the biotech value chain, bringing the company one step closer to commercialization and the customer—the ideal most investors look for. Backward integration often is a concern to investors and thus lowers company value.

Example
In the case of the Arris/Sequana merger, Arris was backward integrated with the purchase of Sequana's combinatorial chemistry capabilities. Although the merger made logical sense to both entities, the market reacted negatively to the combination, and the new company created by the merger, Axys, lost nearly half its value the month after the merger was announced. The marketplace and the financial analysts viewed the backward integration as a negative move away from the end customer.

It is important to remember that the guidelines outlined in this chapter are based on the analysis of six cases and therefore are inherently limited in their instructive scope. They are meant to provide quick "sanity checks" for managers considering an M&A transaction.

NOTES

1. We identified over two hundred public biotechnology companies by studying multiple biotech indexes, Merrill Lynch's Bio-Stats, *Signals Magazine,* and other similar sources to select a sample of firms for financial analysis. These companies were intended to represent a cross section of biotechnology firms, including therapeutic, delivery, tool, and platform companies. We then used CompuStat, DJInteractive, and Yahoo Finance to download historical and current financial data for each company on an annual basis from 1995 forward. We also pulled quarterly data on each company for 2001 and 2002. The data set that resulted comprised 177 public biotech companies with sufficient public information for analysis with data current through the third quarter of 2002. End-of-year numbers were not available at the time of analysis. Cash and short-term investments were used together to estimate the current cash position of the companies analyzed. Burn rates were estimated by taking a company's net income and adding back depreciation and amortization (D&A), as these were noncash items and should not be included in the burn rate. We then calculated a survivability index by dividing cash and short-term investments by the quarterly burn rate in order

to estimate the numbers of quarters remaining for the company. Finally, we translated results to the number of months remaining.

REFERENCES

Applera Corporation. 2002. 10-K filed with the Securities and Exchange Commission on September 27.

Burrill, Steven G. 2003. 17th Annual Report on the Industry. Burrill and Company, San Francisco, Calif. Can be ordered at www.burrilland _co.com.

Copithorne, C., and M. I. Somaiya. 1999. "Gilead Sciences/NeXstar Pharmaceuticals Company Report." Prudential Securities Research Report, April 6.

EquiDesk Database. CommScan EquiDesk/Capital Data. www.comscan .com. Access date April 2002.

King, M., and E. Tenthoff. 1999. "Gilead Sciences, Inc." Biotechnology Research Report. BancBoston Roberston Stephens, April 14.

Longman, R. 1999. "Racing Biotech's Commoditization Clock." *In Vivo,* April 1, p. 46. (A#1999800087 can be ordered at http://archive .windhover.com .content/main.cspx1997).

———. 1997. "Biotech Sidesteps Consolidation." *In Vivo,* December 1, p. 41. (A#1997800241 can be ordered at http://archive.windhover.com .content/main.cspx1997).

———. 2002. "Terra Infirma: Pharma Dealmaking 2001." *In Vivo,* February 1, p. 16. (A#2002800034 can be ordered at http://archive .windhover.com .content/main.cspx1997).

MedImmune. 1999. 10K for 1998. Filed with the Security and Exchange Commission October 21.

Thompson Financial SDC Platinum Database. Thompson.com/financial/Fi _investbank.jsp. Accessed April 2002.

PART III

THE HOME STRETCH:
COMMERCIALIZING
AND CAPTURING
THE VALUE OF THE
INNOVATION

A friend who is a scientist and an entrepreneur used to say, "Once you see the first MBA or engineer in your company, it is time to run." He was, of course, making reference to the unavoidable change of culture that occurs when a biotechnology company transitions from being driven by science to being driven by commercial concerns, wherein processes, strict milestones, budgets, and—the most hated phrase to a scientist—*product prioritizations* seem to dominate the company's vocabulary.

Traditionally, biotechnology companies grow in three sequential stages: (1) the trial-and-error (science-driven) stage, (2) the focus (commercial-driven) stage, and (3) the pipeline diversification (market-driven) stage. The trial-and-error period is an experimental stage when the capabilities of the technology are explored. At its essence, this period is chaotic, unfocused, and highly innovative. Scientists are the decisionmakers, and the chief executive officer (CEO) is often a scientist. Processes are not yet established but are crafted "on the go." Marketing and manufacturing are not yet part of the strategies. At this point, the company is usually funded by angel investors and government grants and perhaps by early-stage venture capital (VC).

The focus period is usually shepherded by VC investors and is signaled by a change of management and the establishment of commercial functions. Often, this is the time when the company prepares to go public.

This is also the most traumatic period for the organization, as scientists hand the reins to seasoned business managers. Organizational tensions abound. Furthermore, the difficult decisions on whether to build or outsource commercial capabilities are made during the focus stage. Some of the struggles that companies undergo in this period are captured in Chapter 6, "Biologics Manufacturing: The Make or Buy Decision." At this stage, companies are at risk of becoming excessively focused, putting their long-term sustainability in danger (see Part II). More often than not, firms are sold at this point.

The third stage is the diversification period, which is triggered in reaction to the company's excessive focus and is signaled again with a change of management. The company forcefully pursues pipeline diversification in this period. By now, the firm has gone public, and its major concern is maintaining the high growth rates expected by the shareholders.

This sequential, stepwise evolution may well lie at the core of most of the problems faced by a growing biotechnology company. The disconnected and shifting strategies (trial-and-error, focus, diversification) trigger instabilities and inefficiencies. A less sequential, more integrated and seamless approach is imperative.

Clearly, a period of trial and error is essential to promote creativity and innovation. But commercial functions can still be integrated without disrupting the process of scientific innovation. Early integration not only lessens organizational disruptions, it also maximizes the profitability of the inventions. Small companies obviously cannot afford to deploy full commercial capabilities. Nevertheless, as described in Chapters 8 and 10, "The Role of Marketing" and "The Forgotten Issue: Reimbursement in Biotechnology," even minimum investments (or simply being attentive to commercialization issues) during the early stages can result in significant benefits to the company. Being cognizant of commercialization issues empowers and bonds the scientists to the business processes, minimizing organizational disruptions later on. In addition, there will be more time to creatively plan the value chain and consider all possible approaches to manufacturing and marketing.

The hectic pace and multiple pressures of building the core research capabilities of a biotechnology company usually result in a disregard for the planning of the commercialization stage. Indeed, commercialization decisions often become afterthoughts. An ingrained inertia then leads to the use of "industry standards" as the major consideration in commercial decisions. In this part of the book, we explore three alternative processes that represent nonstandard opportunities for biotechnology entrants. In Chapter 7, "Pharming Factories," we consider the possibility of manufacturing proteins in animals and plants instead of building brick-and-mortar manufacturing plants. In Chapter 9, "To DTC or Not to DTC," we explore direct-to-consumer marketing as a way to increase the marketing efficiencies for biotechnology. Finally, we take on the payers' function. This topic is commonly considered the sole domain of pharmaceutical companies. However, as explained in Chapter 10, a proactive approach to reimbursement can have a strong impact on the biotech company's profitability.

Chapter 6

BIOLOGICS MANUFACTURING: THE MAKE OR BUY DECISION

Anurag Bagaria

The biotechnology industry is accelerating efforts to discover more protein-based therapeutics (biologics) because of the commercial success to date of over 140 approved biologics. As more molecules are developed, the need for large-scale manufacturing intensifies. Most biotechnology companies have limited resources and must decide whether to devote those resources to building their own manufacturing plants—the "make" option—or to outsourcing the manufacturing to partners or contract manufacturing organizations (CMOs)—the "buy" option. In fact, determining whether to make or buy a manufacturing facility is one of the major decision points in the development of a biotech firm. This chapter addresses the current status of biomanufacturing and presents a strategic rationale and framework to guide the make or buy decision.

The more than 140 approved biologics in the market generated over $30 billion in revenues in 2002, compared to $12 billion just three years earlier (Walsh 2003). Global biopharmaceutical sales in 2002 grew 34 percent over 2001, compared to an 8 percent growth in overall pharmaceutical sales, and they are predicted to grow at 25 percent by the year 2007 (Comer, Stone, and Walsh 2003). This unprecedented growth is a direct result of the increasing research and development (R&D) investment of the biotech industry. The U.S. biotech industry spent $20.5 billion on R&D in 2002. Typically, moreover, 42 percent of revenues are reinvested in R&D (DiLorenzo 2003)—a higher level of investment than in any other major U.S. industry. Biologics that are in the development pipeline consist of approximately five hundred candidates in the clinic and over a thousand in early-phase development, with the majority of this development happening in small or midsize biotech firms.

An increasing proportion of research on biotherapeutics today is focused on complex proteins, such as monoclonal antibodies and fusion proteins. Most of the first generation of biotech drugs were agonists—drugs capable of combining with receptors to initiate a catalytic biological activity. But monoclonal antibodies and fusion proteins are antagonists—agents that oppose the action of other proteins by binding to them in a roughly one-to-one ratio. Consequently, antagonist drugs require more protein by weight than agonist drugs. For example, a billion dollars' worth of erythropoetin (EPO) (an agonist) requires approximately 0.5 kilograms of protein whereas a billion dollars' worth of Enbrel (an antagonist) requires approximately 200 kilograms of protein (Chovav et al. 2003). Moreover, the growing focus on therapeutics for chronic diseases such as diabetes, arthritis, and cancer means that drugs are administered over longer periods of time, again increasing the quantities required for these drugs.

THE IMPORTANCE OF MANUFACTURING CAPACITY

This rapid success in biopharmaceuticals has sent the industry scurrying to develop manufacturing capacity. The unexpected success of the drug Enbrel, from Immunex, in conjunction with a flurry of development activity spurred by huge investments in the biotech industry in 1999 and 2000, set off alarms that alerted the industry to the impending shortfall in biomanufacturing capacity. These

developments motivated both drug developers and contract manufacturers to invest in additional manufacturing capacity, especially
mammalian cell culture capacity (Figures 6.1 and 6.2).

Though nobody can accurately predict whether the increase in
supply will meet the rising demand, most industry experts feel there
is adequate capacity to suffice until 2006 or 2007. However, as
Figures 6.1 and 6.2 show, 81 percent of predicted manufacturing
capacity in 2006 will be captive—that is, held by the matured biopharmaceutical drug developers—and thus will be inaccessible to
the hundreds of small biotech firms performing substantial R&D in
the industry. This concentration of capacity will probably create
pockets of both undercapacity and overcapacity, which will necessitate partnerships, alliances, or mergers. One example of such an

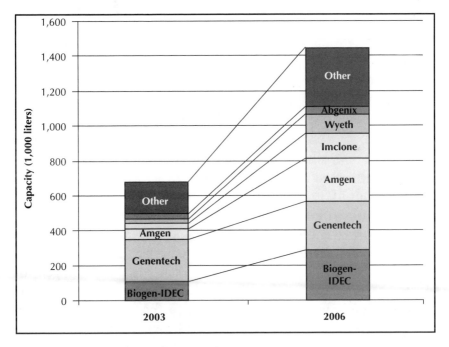

Figure 6.1   Captive Manufacturing Capacity

Sources: Data from company interviews and Chovav et al. 2003.

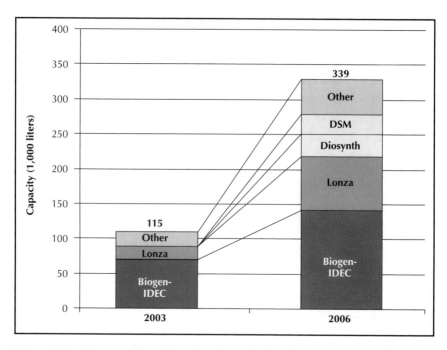

Figure 6.2   Contract Manufacturing Organization (CMO) Capacity

Sources: Data from company interviews and Chovav et al. 2003.

alliance is the deal struck between IDEC and Genentech for the
drug Rituxan. IDEC was forced to find a partner primarily because
of its lack of manufacturing capacity, and in the process, it relin-
quished 70 percent of the drug's revenues to Genentech (Gajilan
2003). Yet another example in which manufacturing capacity was
the key driver of the partnership was Elan's deal with Biogen for
Antegren. And like IDEC, Elan lost a substantial portion of the
drug's revenues, with 50 percent going to Biogen (Griffith 2003).
These deals highlight the importance of manufacturing planning;
they also cause one to wonder how manufacturing was omitted
from the product development plan.

WHY MANUFACTURING IS NEGLECTED

During the biotech boom of the late 1980s and early 1990s, most biotechnology companies were still in the early stages of product development. Investors and Wall Street analysts, knowing no better, valued their portfolio companies by focusing on the product pipelines. By doing so, investors sent a clear signal to company managers that the depth and breadth of the pipeline and clinical progress drove valuations. Management responded by pouring resources, both cash and talent, into their product pipelines, and they advanced their products through the clinic while neglecting manufacturing.

Many executives see manufacturing (if they see it at all) as a downstream issue and tend to focus their time on more immediate tasks. Biotech managers have competing needs for scarce resources, so the tasks of funding research, developing and protecting intellectual property, launching clinical programs, working with the Food and Drug Administration (FDA), raising cash, recruiting talent, making deals, and signing partnership agreements are all higher priorities because these activities are understood by Wall Street. By now, however, the industry has learned that manufacturing is no less important or costly than clinical programs. Manufacturing capabilities and capacity, along with clinical programs, should be important valuation drivers because even with an FDA approval, a company cannot sell what it cannot make.

Industry leaders are largely responsible for ignoring manufacturing and for failing to teach the market about its importance. It could be argued that managers tend to focus on short-term concerns, relegating longer-term issues such as manufacturing to their successors.

The widely known Enbrel shortage as well as other manufacturing-related product delays have highlighted the importance of adequately planning for biomanufacturing capacity (as discussed in Chapter 8, "The Role of Marketing"). The drug had a sales potential of $1.5 billion, but the underestimation of demand and insufficient manufacturing capacity left Enbrel's revenues hovering around $800 million in 2002 (Chovav et al. 2003).

Thus, managers need to develop effective manufacturing strategies to assure that capacity is available when needed. This goal can be accomplished through direct build, partnerships, or contract manufacturing. Successful managers will have to develop the business plans and governance capabilities to integrate multiple processes—manufacturing development, process R&D, tech transfer, scale-up, quality and regulatory assurance—and to do this seamlessly with other product development activities.

In the early days, all biotech firms wanted to be fully integrated pharmaceutical companies (FIPCOs), and they started to build plants for their lead compounds. Because building, staffing, and validating a plant costs a lot of money and takes a lot of time, many companies did not survive this strategy when their products failed in the clinic (e.g., Synergen, in Boulder, Colorado). However, the biologics contract manufacturing industry was in its infancy at that time, so these companies did not have much of a choice in outsourcing manufacturing. As a consequence, the virtual integrated pharmaceutical company (VIPCO) concept became the new strategy, and thus, the trend shifted toward buying capacity. This shift from in-house manufacturing to outsourced manufacturing triggered the growth of the contracting business. Given that the outsourcing business is growing and maturing, companies now have a useful and realistic option in outsourcing, which means the make or buy decision is decidedly relevant in the current environment.

Today, biopharmaceutical companies have three viable options for manufacturing their products: (1) they can contract production with a third-party contract manufacturer, (2) they can build a manufacturing facility, or (3) they can license manufacturing and marketing rights for the product to another company in exchange for up-front cash payments and royalties on future sales.

The outlicensing option is one that many small biotechs rely on, as many lack both the manufacturing and marketing infrastructure required to commercialize products. However, as seen from the Rituxan example, this is an expensive route, for the developer of the drug squanders an enormous amount of future revenues to the licensing partner. Even though it is an expensive option, many small biotech companies hope to partner with Big Pharma/biotech, and thus they neglect the manufacturing plan. Of course, they have to buy the manufacturing for clinical trial material, but they postpone the decisions for commercial manufacturing.

The smaller biotechs need to recognize that outlicensing a product to Big Pharma is not a trivial process. Big Pharma's in-house R&D often shelves any product that does not have a market potential of around $500 million; thus, it will certainly not buy a compound with low sales potential from a small biotech. In addition, even biotechs are spinning off smaller companies that can work on individual products with smaller market potential. For example, Tercica was spun off from Genentech to develop and commercialize a growth factor. Similarly, Paion was spun off from Schering AG to pursue opportunities in treating strokes. Outlicensing is also challenging because of the competition and complex valuation methods used for biotech products. Given the large amount of compounds in clinical trials, Big Pharma has innumerable choices to select from, further complicating the inlicensing process. As Big Pharma gets bombarded with proposals for new products, it is turning many down because it does not have the time and resources to evaluate all of them. Consequently, only a small number of these compounds will actually be bought, and all the other compounds with smaller market values will have to make it on their own.

To prevent delays in the development and commercialization of their products while still conserving vital capital, companies must carefully analyze their technical, strategic, and financial options and select the best alternative. For production of early-stage clinical trial materials, outsourcing manufacturing often provides the faster route to clinical trials and delays the need to invest capital in "bricks and mortar." However, for many companies, as their cost of capital becomes more favorable and expected returns on investment go up, construction of a manufacturing facility may make sense. This is a difficult decision to make, with numerous alternatives and many more aspects to evaluate. The final make or buy decision should be based on a cost-benefit analysis in conjunction with an evaluation of the company's resources and objectives.

A STRATEGIC ANALYSIS OF THE MAKE OR BUY DECISION

The following sections present a discussion of four important strategic questions that management should try to answer when deciding whether to make or buy manufacturing capacity.

WHAT IS THE VISION FOR THE FUTURE OF THE COMPANY?

To answer this question, a biotech company's management might consider the following factors:

1. Whether the company will become a fully integrated biopharmaceutical company
2. Whether the company will outlicense its products to bigger pharma/biotech
3. Whether the company's vision is to be acquired by bigger pharma/biotech

A company's vision is often decided (or should be decided) by the founders of the firm. Though the answer to the preceding question might evolve over time given various circumstances, it should result from discussions with upper management or the board of directors. If the company is planning to outlicense or be acquired, then it does not make sense to build a manufacturing facility because the inlicenser or acquirer will provide this utility. At most, the company can construct a small-scale pilot plant and build process development expertise to support its clinical development activities. For a company that envisions becoming an integrated player, the make or buy decision will hinge on the cost-benefit analysis and the other strategic aspects discussed in the passages that follow.

HOW STRONG AND DEEP IS THE PIPELINE, AND ARE THERE OTHER POTENTIAL PRODUCTS THAT WILL BE MANUFACTURED USING THE SAME TECHNOLOGY?

An evaluation of the company's pipeline is critical to the make or buy decision. If the pipeline is weak or shallow or if products in the pipeline use different technologies—mammalian versus microbial fermentation, small versus large molecules—then building a manufacturing facility is a very risky proposition. An alternative for a company in this situation is to initially outsource its needs and then, depending on the success of the product, build a facility and bring the manufacturing back in-house. However, if the company has a strong pipeline of products requiring a similar technology, then vertical integration can be a viable option.

WHERE IS THE PRODUCT ALONG THE PRODUCT LIFE-CYCLE CURVE?

There are two types of uncertainty in the life cycle of a product: development uncertainty and market uncertainty. As the product matures and passes through both the development and market uncertainties, the risk of buying exceeds the risk of making. But because the make decision has to be arrived at during the period of development uncertainty, this strategy carries tremendous risk, for the chances of product and market failure are high. In the early stages of a company, some firms might decide to build pilot plants and process development expertise for better control during the development phase; others might choose to buy their initial needs and then make the product when the uncertainties have been resolved. As the firm matures and builds a strong portfolio of products on the market and in the pipeline, it can allocate resources toward the pilot plant and large-scale commercial manufacturing. One analytical tool that can be useful and is increasingly popular in decisionmaking contexts is real options. A growing number of companies (e.g., Merck) are using this tool to manage their pipelines; others (e.g., Wyeth Biopharma) are evaluating its usage in manufacturing planning.

WHAT ROLE WILL THE MANUFACTURING STRATEGY PLAY WITHIN THE OVERALL BUSINESS STRATEGY?

The answer to this question will drive the amount of resources the company has to devote to manufacturing operations without hurting other business functions (for instance, marketing/sales and R&D). The question is best answered by using the framework described by S. C. Wheelwright and R. H. Hayes in "Competing through Manufacturing" (Wheelwright and Hayes 1985). The authors describe four roles a company's manufacturing strategy can play; these roles can be viewed as stages of development along a continuum.

At one end of the continuum is the "internally neutral" stage, wherein the managers seek to minimize the "negative" impact of manufacturing (on the company's culture and resources) and believe that manufacturing does not affect the company's overall competitive position (this thinking is prevalent in most early-stage biotech companies). At the other end of the continuum is the "externally supportive" stage, wherein the competitive strategy of the firm rests, to a significant degree, on its manufacturing capability. Genentech is an example of a biomanufacturing company at the externally supportive stage with regard to its manufacturing department. The company continuously anticipates the potential of new manufacturing practices and technologies and seeks to acquire expertise in them long before their implications are fully apparent. Unlike its competitors, Genentech also places equal emphasis on structural activities (involving the building and equipment) and infrastructural activities (involving management policies) as potential sources of ongoing improvement and competitive advantage.

A biomanufacturing supportive strategy is expensive to implement. A collaborative culture has to be established to encourage the interactive development of business, research, manufacturing, and other functional strategies. The process and product have to be codeveloped, which means that scientists and engineers have to work together—not an easy task. As a company's managers analyze the make or buy decision, they should consider the implications of the decision on their resources.

A strategic analysis of this nature helps a company align its manufacturing strategy with its business strategy. Once these strategies are aligned, the company should perform a cost-benefit analysis to weigh the resources required against the resources available to make short-term and long-term decisions.

COST-BENEFIT ANALYSIS

THE BENEFIT SIDE OF THE EQUATION

Various benefits are associated with using the market for biologics manufacturing. The benefits, both tangible and intangible, include economies of scale, lower agency and influence costs, learning economies, increased speed to market, risk reduction, and converting fixed costs to variable costs.

Economies of Scale
Both the capital expense of building the facility and the operating costs of using it do not scale linearly with fermentation volume. Thus, as the scale increases, the cost per unit decreases substantially. Because CMOs can aggregate the needs of many firms, they enjoy economies of scale and can offer lower costs to each of their clients.

Lower Agency Costs
By using the market, a firm can reduce or eliminate agency costs. These are costs associated with slacking by employees and the administrative efforts to deter such slacking. The cost of slacking in biomanufacturing is primarily associated with the talent shortfall. The industry faces a looming shortage of highly trained people to design, build, and operate facilities. Experienced process development scientists and engineers, validation engineers, quality assurance personnel, and plant managers are also in short supply. As a result, companies must offer powerful incentives to reduce the likelihood that their employees will search for better opportunities. Additionally, though the Biogens, Genentechs, and Amgens of the

world can attract, train, and retain the required personnel at reasonable costs, it is very expensive and difficult for smaller companies to do so. Furthermore, as demand increases with the addition of new capacity, attracting and retaining employees will become that much more expensive and challenging.

Lower Influence Costs

Influence costs are the costs of activities designed to influence internal capital markets. They include not only the direct costs involved in influence activities (e.g., time spent by a division manager in lobbying central management) but also the costs of bad decisions that arise from such activities. These costs are especially high when internal capital is scarce, as all divisions work hard to gain a higher share of the capital available. In biotech firms with internal manufacturing divisions, this cost can be exceptionally high because capital is often scarce, especially today, and there is an allocation disconnect between R&D and manufacturing. This is particularly true in the case of small biotech firms, for manufacturing can be a distraction from R&D, their core competency.

Learning Economies

Learning economies are very valuable in the new and developing field of biomanufacturing. Contract manufacturing organizations can exploit their steep learning curves, a consequence of working on different products for their various clients, to reduce costs further. The talent shortfall in the industry amplifies the learning economies. As the demand for talent far outstrips the supply of talent, there is a substantial turnover of key personnel in the industry. Compared to midsize and small biotech companies, the established CMOs are better positioned to train talent (due to an already established talent pool) and retain talent (due to a lower risk and a wider variety of projects). Thus, they provide the expertise required at lower costs. Additionally, some contract organizations possess proprietary information or patents that enable them to produce at lower costs as well; an example is Lonza's proprietary gene expression system.

Increased Speed to Market

Outsourcing manufacturing to a CMO can help a sponsor substantially reduce the time to market. Though finding, qualifying, and choosing a contractor is a time consuming process, it is faster than building a facility from scratch. Identifying the right contractor is only the first step and can take up to six months. The steps following identification can take anywhere from twelve to eighteen months and include contract negotiation, equipment checking, validation, technology transfer, shakedown runs, and finally the real batch manufacturing. When commercial manufacturing is outsourced, there is an added step of regulatory filing that takes around twelve months (Figure 6.3).

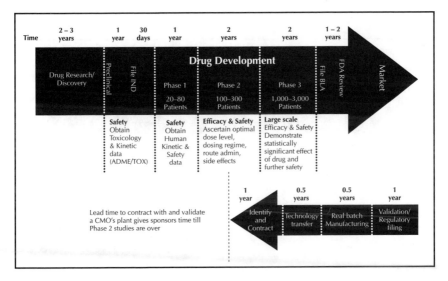

Figure 6.3   Proposed Outsourcing Time Line

Source: Company interviews.
Note: IND = investigational new drug; BLA = biologics licensing application; CMO = contract manufacturing outsourcing; ADME/TOX = adsorption, distribution, metabolism, excretion, and toxicity testing.

Risk Reduction

Outsourcing to a CMO substantially reduces the financial risk for the sponsors by (1) delaying decisions on capital expenditure, and (2) delaying decisions on capacity requirements.

In the first instance, risk can be reduced by delaying (or abandoning) decisions on capital expenditure in a project that has a very low probability of success. It takes approximately four years to go "from dirt to drug" (building a small-scale manufacturing plant for clinical material takes around two and a half to three years), and thus, the construction of the manufacturing plant would have to start before phase II data are available (Figure 6.4). During phase II, in addition to the resources allocated to clinical development, resources need to be devoted to developing a robust manufacturing process for the product. Process development during this period is extremely important because the Food and Drug Administration (FDA) stipulates that there should not be any significant changes in

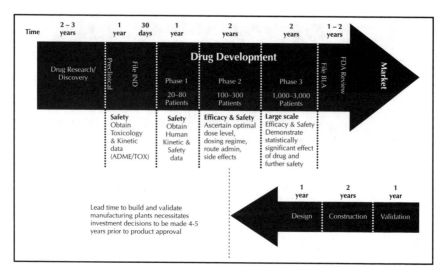

Figure 6.4   Proposed Construction Time Line for a New Manufacturing Plant

Source: Company interviews.
Note: IND = investigational new drug; BLA = biologics licensing application; ADME/TOX = adsorption, distribution, metabolism, excretion, and toxicity testing.

the manufacturing process between phase III and commercial manufacturing (see the Genentech minicase).

In the second instance, risk can be reduced by delaying decisions regarding capacity in an early stage in a product's clinical development—a stage at which little is known about the drug. Among other objectives, phase II studies are performed to ascertain the optimal dosage level and dosing regime for the drug. These attributes, along with market potential (which is a wild guess during phase II studies), drive the quantity required for the drug. Yet

Minicase: Genentech

Late in 2001, the biologics license application (BLA) for Xoma and Genentech's psoriasis drug Raptiva was delayed after Genentech's phase III product failed bioequivalence tests when compared to Xoma's earlier product. Genentech's modification to the scale-up process resulted in slightly higher serum concentrations than in Xoma's version of the drug (which was used in phases I and II). The FDA was unhappy about the lack of bioequivalence, and the companies were forced to delay the BLA for eighteen months, until December 2002. And we all know how damaging delays can be to a company's bottom line. When there is a process change, the sponsor has to conduct additional clinical trials to prove the bioequivalence of the product after the change. Thus, an allocation of resources toward a Greenfield manufacturing plant, in addition to process and clinical development, can be a huge burden on the product's sponsor. Moreover, the risk in the project is also very high, stemming from the fact that investment decisions and layouts begin in phase II when the probability of success for a biologic is extremely low, at 15 percent. By outsourcing to a contract manufacturing organization, the investment layouts can be postponed to the future, when there will be more data on the success of the product. After establishing product success, if the sponsor does decide to build a manufacturing plant, it will be able to bring the manufacturing back in-house after about two to three years of outsourcing. However, this decision will also depend on the drugs in the company pipeline and the technology required to manufacture them.

another uncertainty is the recovery yield from the manufacturing process, for phase II studies are not necessarily good predictors of the phase II recovery yield. As seen in Figure 6.5, these uncertainties can translate into one to two orders of miscalculation about the capacity required. As we know from the Enbrel example, the upside lost from undercapacity probably outweighs the cost burden of overcapacity. However, if the estimations are completely off or if the product fails, the burden of maintaining a manufacturing plant can drive a company to bankruptcy.

Converting Fixed Cost to Variable Cost
Building a manufacturing facility in any industry is, of course, a huge capital investment. In the world of biologics, the maintenance and operation of the manufacturing facility is also an enormous fixed cost.

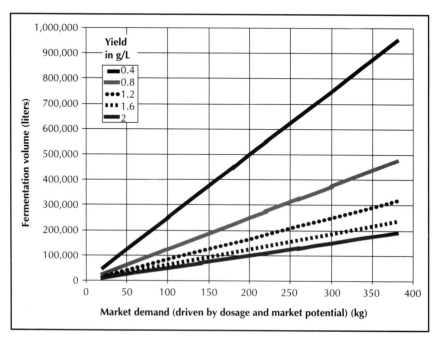

Figure 6.5   Sensitivity of Fermentation Volume Required on Yield, Dosage, and Market Demand

Sources: Graph drawn from data obtained from UBS and the Internet

Building a commercial manufacturing plant with a capacity of 30,000 liters to 40,000 liters costs $300 million to $400 million, and operating costs range from $80 million and $120 million, 50 percent of which is fixed cost at full utilization. These high fixed costs are a result of manpower and maintenance requirements. Expenditures in manpower tend to be fixed in this industry because of high training costs, and maintenance of the facility is required around the year to ensure that the plant is compliant with good manufacturing practices (GMP). Though these costs are not completely avoided when a sponsor outsources manufacturing, a good proportion of them are shared between the various clients of the CMO, and more important, they become variable costs for the sponsor.

THE COST SIDE OF THE EQUATION

Several tangible and intangible costs must be considered as a manager makes the make or buy decision: loss of control, learning loss, coordination costs, transaction costs, and leakage of information.

Loss of Control

Probably the single biggest cost of outsourcing manufacturing is the loss of control. The sponsor has complete control of both the process and the facility when it manufactures the product in-house. By contrast, when the sponsor outsources work, client confidentiality prevents the contract manufacturing organization from revealing specifics of other products manufactured in its facility. Though a certain amount of information has to be revealed to ensure no cross-contamination or biological interactions occur between two different products, other aspects of facility maintenance and/or use might be omitted, even though this could impact the integrity of a biological product. Other control issues are related to regulatory compliance. Because the sponsor does not have full control over the contractors' actions, it has to trust the contractor to take all necessary steps to gain or maintain regulatory compliance. Thus, the sponsor takes a substantial risk, for a formal warning letter from the FDA can increase time to market, hurt sponsor revenues, and disgruntle patients who are dependent on the drug. The importance of this factor is further highlighted in the Bayer's Kogenate minicase. The loss of control is prominent in the process development and capacity allocation decision as well. In an in-house manufacturing/process development facility, the sponsor can continuously reprioritize the various products, depending on the goals of the overall organization. That is not the case when dealing with a contract organization, and the sponsor will face severe penalties when changes in the production-development schedule are made.

Minicase: Bayer's Kogenate FS

The manufacturing plant for Bayer's Kogenate FS (a recombinant therapeutic for the treatment of hemophilia) was in Berkeley, California. Originally constructed in 1993 to manufacture Kogenate, the plant was modified in 1999–2000 to manufacture an improved version, Kogenate FS. Bayer hoped for a smooth transition in the summer of 2000 and was awaiting FDA approval, expected in January 2000. But as industry followers know, the timing of FDA product approvals is very uncertain, and the approval did not come for Bayer until July 2000. On the arrival of the clearance letter, Bayer started ramping up production of the new and improved Kogenate FS. However, in November 2000, the FDA showed up for a periodic inspection—unannounced, as is the agency's normal practice. After rigorous inspections of the plant for six weeks, the FDA issued a formal warning letter to Bayer in February 2001, which put a halt to all manufacturing operations. It took eighteen months and $30 million, as well as the replacement of the plant's top managers and a change in culture (which facilitated more frequent communication between employees in various manufacturing functions) for Bayer to regain FDA compliance and resume product manufacturing. This example illustrates two important points: (1) to maintain or gain FDA compliance, a company must scrutinize, daily and meticulously, every detail of its operation and carefully monitor its high costs, and (2) that if an experienced firm such as Bayer can become negligent in its own, in-house facility, so can any other firm or contract manufacturing organization.

Learning Loss

If a firm outsources its manufacturing, it cannot learn from process development and/or large-scale manufacturing. This might not be a big concern in small-molecule manufacturing, but it is very important in biologics manufacturing. Because a biological product is primarily characterized by the manufacturing process, a great deal is learned about the product itself during drug development. Thus, if a sponsor outsources its manufacturing, it loses not only the opportunity to build manufacturing expertise but also the opportunity to learn more about the product. This learning loss can be expensive during clinical development. The cost of learning loss can be minimized by ensuring that the relationship with the CMO is structured as a partnership with a rich flow of information between partners.

Coordination Costs

Outsourcing demands excellent project management capabilities. A supply delay caused by a missed deadline or a product that fails analytical testing (e.g., a failed batch due to bad quality) leads to substantial lost revenues. And if the drug is in clinical trials, a missed deadline will increase the time to market, again resulting in lost future revenues. Moreover, a late batch from the manufacturer can raise capacity management issues at the sponsor's fill/finish contractor. This in turn can lead to further delays in the release of the final product. Best practices indicate that a sponsor should employ three to four dedicated personnel with quality assurance/quality control (QA/QC), manufacturing, process science, and supply chain experience. Other measures that help reduce this cost include contracts structured to incorporate incentives for keeping schedules, matching business cultures, and fluid communication.

Transaction Costs

Transaction costs include the time and expense involved in negotiating, writing, and enforcing contracts. These costs are exceptionally high when one party acts opportunistically while the other tries to prevent it from doing so. This is probably true when the demand exceeds the supply for capacity and the CMOs spend time and effort to sweeten their deals. An additional reason for high transaction costs is the complexity in the relationship between the sponsor and the CMO. Because both the relationship and the manufacturing process are so complex, the ability to write complete contracts that safeguard each party is limited; further, the temptation for a party to hold up its trading partner is likely to lead to the frequent renegotiation of contracts.

The "hold up" problem in biomanufacturing is actually reversed, as the sponsor cannot hold up the contractor due to investments in relationship-specific assets. Conversely, the contractor can hold up the sponsor because of the criticality of time to market and the long lead times entailed in switching from one contract organization to another due to technology transfer risks. Furthermore, the CMO might have used proprietary knowledge to improve the manufacturing process of the sponsors' product. In such a situation, if the sponsor later decides to transfer the product to an in-house facility or another contract organization, it will either lose the productivity gains from the CMO's proprietary knowledge or it will have to pay royalties for using the knowledge. Thus, the sponsor might be vulnerable to hold up, which results in higher transaction costs and delays in manufacturing. This problem can be particularly serious in biomanufacturing, as precise synchronization is required between the bulk output and capacity availability at the fill/finish contractor.

Yet another cause for the complex relationship—and consequently, high transaction costs—is the negotiation for new intellectual property (IP) coming out of process development. Intellectual property negotiation can be a sore topic of discussion, as both parties will want to patent any new discoveries for the sake of future benefits. The conflict in IP ownership arises because the sponsor pays for the development work, whereas the contractor deploys skills and manpower to actually do the development work. One solution that is prominent in the industry is to leave any new generic process IP with the contractor and transfer any product-specific IP to the sponsor. In any case, this issue should be clarified during the initial negotiations.

Leakage of Information

Loss of confidentiality is a major cost to consider when contemplating outsourcing. Though most contract manufacturing organizations are not in the product development business, precautions should be taken to avoid future conflict by including a noncompete clause in the contract. Comprehensive legal protection is important and necessary when the contractor is a potential competitor. This is the case when integrated biotech/Pharma companies such as GlaxcoSmithKline (GSK), Abbott, and Genentech sell their excess capacity. For example, Amgen has signed a supply contract with Genentech for Enbrel and Genentech is developing Raptiva, a drug that will compete with Enbrel as a treatment for psoriasis; some legal confrontations might ensue.

THE DECISION

Performing the cost-benefit analysis is critical for arriving at the right make or buy decision. The following sections offer examples of what firms of different sizes are doing in this regard.

SMALL BIOTECHS

Most small biotech companies with no product revenue stream have uncertain long-term futures and limited resources and thus choose to buy capacity. This decision is primarily based on the enormous capital expenditure required to build the facility, followed by the huge operation expenses required to run it. Smaller biotech companies are better off buying the capacity (at least for the first product) and delaying the risky investment in infrastructure.

One example of a small biotech that made this decision is Tercica Corporation. Tercica was established in May 2002 with one inlicensed product from Genentech that completed phase III trials. At that point, Tercica was in a better situation than most other small biotechs because it did not need any research or clinical development. Moreover, Tercica would not have to spend any money on process development for its product—an insulin-like growth hormone—as a robust manufacturing process had already been developed by Genentech. Yet even in this circumstance, with only one product and with limited cash (around $70 million from two rounds of financing), Tercica decided to look for a contract manufacturing organization instead of building its own capacity. Like other small biotechs, Tercica based its decision on the lack of funds, the high cost of capital, the unavailability of in-house capacity, and a limited pipeline. The company decided its dollars would be better spent on building sales and marketing infrastructure to take its product to market. In this way, the company also earned the value of the option to build a facility in the long run. In the future, if Tercica's goal is to become a vertically integrated company, it can build a dedicated facility for its successful product. Or as the pipeline develops with a steady revenue stream from one product, it can build a flexible facility for the development and commercialization of its future products.

MIDSIZE BIOTECHS

Midsize biotechs are taking a different approach to manufacturing. These companies generally have a more robust pipeline and have experienced process development activities in addition to a stronger cash position. Although some firms have chosen to outsource their manufacturing needs, others have invested in manufacturing infrastructure. Again, because the financial commitment to build and operate a facility can excessively strain the operations of a firm, even a midsize company needs to approach the make or buy decision with caution.

MedImmune, for example, built a manufacturing facility to make its product Synagis. The company made this decision as a consequence of two important accomplishments: (1) it already had developed a product (RespiGam) that was generating revenue, and (2) it had achieved positive phase II results for Synagis. Thus, the financial risk of the facility was mitigated, and the probability of success for Synagis was much higher. Moreover, MedImmune had a strong portfolio and would potentially need manufacturing in the near future, and it could source expertise from its pilot plant. This strategy had a much lower risk of failure. Going forward, MedImmune will continue to outsource initial manufacturing requirements and build facilities when drug success is much more probable.

By contrast, Human Genome Sciences, Inc. (HGSI) is investing in a large-scale manufacturing facility even though it has no drugs currently approved for marketing. However, HGSI does have a very strong cash position, strong pilot plant operations, and a rich pipeline. Nonetheless, this is a risky strategy because any failure of its products in the clinic would make the facility a huge burden.

LARGE BIOTECHS

Large biotech firms (including Genentech, Biogen, and Genzyme) evolved slowly to become the integrated companies they are today. In the initial phase of their existence, these companies outlicensed the marketing rights for their first products to Big Pharma (e.g., Genentech outlicensed insulin to Lilly, and Biogen outlicensed interferon to Schering Plough), and at the same time, they transferred the task of manufacturing, too. These deals with Big Pharma provided a steady source of revenue from royalty payments for the growing biotech companies, which continued to grow and become financially self-sufficient. At that point, the large biotechs recognized the importance of manufacturing and marketing and started to build in-house expertise. They evolved from research-based companies to vertically integrated biopharmaceutical companies. Today, they boast the best-in-class manufacturing facilities and expertise that other companies hope to get access to through partnerships.

FRAMEWORK AND FINAL RECOMMENDATION

A company should build manufacturing infrastructure only if its strategic goal is to become a vertically integrated firm, if it has a robust pipeline that requires similar manufacturing technology, if the manufacturing strategy is internally supportive to the business strategy, and if the benefits of making outweigh the costs of buying. As a firm evolves and grows, these different decision parameters constantly change; hence, the decision requires periodic reevaluation (Figure 6.6).

Another widely used rule of thumb that companies employ to justify a manufacturing facility is to have at least one drug approved for the market and a couple more with good potential in the pipeline. In this situation, the financial risks are mitigated by the assured revenue stream from the approved product, and the approach makes strategic sense due to the potential in the pipeline.

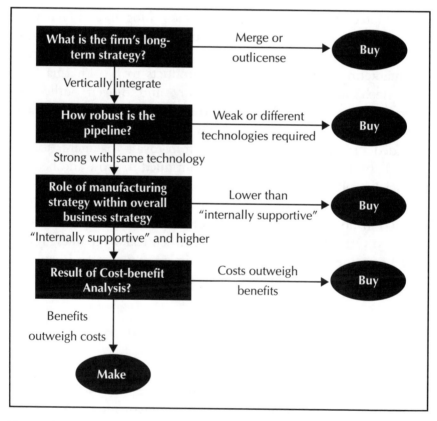

Figure 6.6   Decisionmaking Framework

The make or buy decision is a very difficult one and has long-term strategic implications that companies must consider. The right decision will vary not only among different companies but also within a given company depending on what stage it is at in its life cycle. Strategic and cost-benefit analyses should guide the decision of whether to buy or build. These analyses are essential to increasing the commercialization efficiencies that will benefit the biotech industry and society at large.

REFERENCES

Chovav, M., M. Wales, D. De Bruin, A. Samimv, G. Meacham, K. Kim, and D. Farhadi. 2003. "The State of Biomanufacturing." UBS Report, June, 1–52.

Comer, A., M. Stone, and P. Walsh. 2003. "Lonza." HSBC Report, June, 1–36.

DiLorenzo, F. 2003. "Industry Surveys: Biotechnology." *Standard & Poor's,* May, 1–44.

Gajilan, A. T. 2003. "The FSB 100: The Graduate." *Fortune.* Available at http://www.fortune.com/fortune/smallbusiness/smallcaps/articles/0,15114,457908,00.html. Accessed June 2003.

Griffith, T. 2003. "Biogen-Idec Counting on New MS Drug." *CBS Marketwatch.* Available at http://cbs.marketwatch.com/news/story.asp?guid=%7BAAF5CB0A-AB1D-4628-99C2-08122114A531%7D&siteid=google&dist=google. Accessed June 2003.

Walsh, G. 2003. "Biopharmaceutical Benchmarks." *Nature Biotechnology* 21: 865–870.

Wheelwright, S. C., and R. H. Hayes. 1985. "Competing through Manufacturing." *Harvard Business Review,* January–February, 1–12.

Chapter 7

PHARMING FACTORIES

*Andrew Ingley, James Pavlik, and
Ty Smith*

The pharmaceutical and biotechnology industries are intensely interested in the structure, function, production, and potential use of proteins as therapies to treat diseases. Much of the success within the biotech industry derives from the development of replicated human proteins to treat diseases that occur when the body makes too little of an important protein or when the presence of unusually large amounts of a protein generate a disease. A sample of approved biologicals is shown in Table 7.1. Many biopharmaceuticals target chronic diseases such as cancer and immune disorders, which typically require relatively large and repeated doses. As a result of the expected increase in high-volume biologic therapeutics on the market, many industry analysts have predicted significant shortages in the supply of biologics manufacturing capacity.

As mentioned in the previous chapter, contract manufacturing could be an option when dealing with the increased manufacturing capacity. However, most contract biomanufacturers are approaching 100 percent capacity utilization rates for both mammalian cell culture and microbial fermentation capacity (Fox, Khoury, and Sopchak 2001). It is estimated that the industry will require a five- to sixfold increase in cell culture capacity as biologics in the pipeline reach commercialization (Andersson and Mynahan 2001: 1–5).

Table 7.1
Sample of Major Approved Biotechnology Products

Product	Company	Indication
Nonantibody Proteins		
Actimmune	InterMune	Chronic granulomatous disease
Activase	Genentech	Acute myocardial infarction
Aranesp	Amgen	Anemia
AVINZA	Ligand Pharmaceuticals and Elan Corp. PLC	Pain
Avonex	Biogen	Relapsing multiple sclerosis
BeneFIX	Genetics Institute	Hemophilia B
Betaseron	Chiron/Shering AG	Relapsing, remitting multiple sclerosis
Cerzyme	Genzyme	Gaucher's disease
Crosseal	OMRIX	Adjunct to hemostasis
Eligard	Atrix Laboratories and Sanofi-Synthelabo	Prostate cancer
Epogen	Amgen	Anemia (Chronic Renal Failure)
Elitek	Sanofi-Synthelabo	Pediatric chemotherapy
FluMist	MedImmune	Vaccine, influenza virus
FortaFlex	Organogenesis, Inc., and Biomet, Inc.	Cuff repair
FORTEO	Eli Lilly	Osteoporosis
Hepsera	Gilead	Hepatitis B
Intron A	Shering-Plough	Hairy cell leukemia
INFUSE	Wyeth and Medtronic Sofamor Danek	Spinal surgery
Kineret	Amgen	Rheumatoid arthritis
Leukine	Immunex	Autologous bone marrow transplant
Natrecor	Scios	Congestive heart failure
Neulasta	Amgen	Infection as manifested by febrile neutropenia in nonmyeloid cancer
Neumega	Genetics Institute	Thrombocytopenia
Neupogen	Amgen	Neutropenia
NovoSeven	NovoNordisk	Hemophilia A bleeding
Nutropin	Genentech	HGH deficiency
Ontak	Ligand Pharmaceuticals	Cutaneous T-cell lymphoma
Orfadin	Swedish Orphan	Hereditary tyrosinemia type
Pediarix	GSK	Vaccine, diphteria, tetanus, hepatitis B, and polio pertussis
PEG Intron	Shering-Plough	Hepatitis C
Pegasys	Roche and Inhale Therapeutics, Inc.	Hepatitis C
Procrit	Amgen/Johnson and Johnson	Anemia (oncology)
Proleukin	Chiron	Renal cell carcinoma

Table 7.1 (continued)

Product	Company	Indication
Pulmozyme	Genentech	Respiratory infections (cystic fibrosis)
ReFacto	Genetics Institute	Hemophilia A bleeding
Rebif	Serono SA and Pfizer Inc.	Multiple sclerosis
Remodulin	United Therapeutics Corporation	Treatment of pulmonary arterial hypertension
RESTASIS	Allergan, Inc.	Ocular inflammation
Roferon-A	Roche	Hairy cell leukemia
SecreFlo	Repligen Corporation	Pancreatic assessment
Xigris	Lilly	Sepsis
Xyrem	Orphan Medical, Inc.	Cataplexy associated with narcolepsy
Monoclonal Antibodies/Fusion Proteins		
Campath	ILEX Oncology/Berlex	Chronic lymphocytic leukemia
Enbrel	Immunex	Rheumatoid arthritis
Herceptin	Genentech	Metastatic breast cancer
Humira	Cambridge Antibody and Abbott	Rheumatoid arthritis
Mitozytrex	SuperGen, Inc.	Stomach cancer
Mylotarg	Wyeth-Ayerst	Relapsed acute myeloid leukemia
OKT-3	Johnson and Johnson	Acute kidney transplant rejection
Remicade	Johnson and Johnson	Rheumatoid arthritis, Crohn's disease
Reopro	Lilly	Anti–blood clotting agent (PCTA)
Rituxan	Genentech/IDEC	Relapsed, refractory non-Hodgkin's lymphoma
Synagis	MedImmune	Respiratory syncytial virus infections
Xolair	Genentech	Asthma
Zenapax	Protein Design Labs/Roche	Acute kidney transplant rejection
Zevalin	IDEC	non-Hodgkin's lymphoma

Source: Data obtained from http://www.bio.org/speeches/pubs/er/approveddrugs.asp June 2004.

Even if more capacity could be built, the industry would face high costs associated with biomanufacturing. In spite of technological advancements in cell culture methods, manufacturing complex proteins with yeast, bacterial, or mammalian cell culture systems remains expensive. For example, the estimated costs for constructing a large-scale biologics manufacturing facility with a bioreactor capacity of 100,000 liters range from approximately $200 million to $400 million. Furthermore, the construction and validation of these facilities typically require long lead times of approximately four to five years (Figure 7.1). As a result, drug developers have

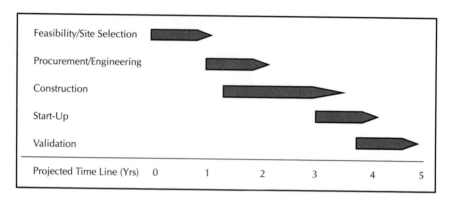

Figure 7.1   Biologics Manufacturing Facility Time Line

Sources: Data from multiple sources and Fox, Khoury, and Sopchak 2001.

often been unable to economically justify the use of biopharmaceuticals (Ginsberg, Bhatia, and McMinn 2002: 1–23).

The substantial amount of time required to construct large-scale facilities has imposed significant business and financial risks on pharmaceutical firms. Companies are generally forced to invest in capacity when their products are in early phase III clinical trials so the facilities are ready when the drugs gain approval from the Food and Drug Administration (FDA). Given the uncertainty related to the outcome of phase III clinical trials, pharmaceutical companies are faced with a difficult decision that often leads to overbuilding or underbuilding capacity. In the past, the primary risk was typically overbuilding capacity. Several facilities that were built for drug manufacturing sat idle after companies failed to win FDA approval and squandered the substantial up-front investments. As demonstrated in the Immunex example of the previous chapter, companies currently face tremendous risks of underbuilding capacity, which can generate significant losses in terms of time to market and potentially hundreds of millions of dollars in sales.

In recent years, transgenic technology emerged as an attractive alternative for protein production primarily because of its ability to produce complex proteins at higher expression levels and in greater volumes than traditional cell culture methods, thereby allowing for lower cost production and more rapid scaling of capacity. In certain instances, in which very large amounts of material are required for therapy, the use of transgenic animals may be the only feasible current technology. Transgenic technologies can produce proteins at levels between 3 to 10 grams per liter, with significant variance depending on the type of protein being produced. If we assumed an expression level of 5 grams per liter, then 100 kilograms could be produced with approximately 20 goats. This figure compares favorably to the optimum yield of approximately 1 gram per liter that can be expected from the best conventional cell culture systems.

Several companies are currently pursuing human clinical trials using proteins derived from the milk or urine of transgenic animals. This process has been termed *pharming* due to its close ties to both the pharmaceutical and agricultural industries. Although pharming promises to offer significant advantages in terms of production cost and scalability for complex proteins, there remain major issues with the technology surrounding animal husbandry, societal acceptance of transgenically derived proteins and transgenic organisms, and regulatory approval.

TRANSGENIC PROTEIN PRODUCTION

Plant and animal pharming can be used for manufacturing drugs. The most common type of animal pharming uses the mammary gland for drug production. Milk has high protein levels and is produced in large quantities, so it is a good vehicle for directed protein expression. Transgenic protein production in milk involves the insertion of a human gene into a fertilized egg or an early-stage embryo. There are many techniques for integrating the deoxyribonucleic acid (DNA) into the embryo, and this is a critical step in ensuring the successful transfer of the gene to the offspring. The embryo is inserted into a surrogate mother, and when the genes are successfully integrated, they are passed on to the offspring and the human protein is expressed in the milk of the female offspring. The first-generation transgenic animals are called founders, and the human protein is secreted in their milk. This milk is collected, and the expressed protein is then extracted and purified. The purification process for transgenic proteins is readily integrated with machinery typically used in purification for cell culture production.

Transgenic technology enables the production of complex proteins. Many proteins now in development represent manufacturing challenges due to complicated folding requirements, posttranslational steps, and other factors. Nonsecreted proteins, fusion proteins, and monoclonal antibodies can be difficult to produce in many of the standard manufacturing technologies but can be produced using transgenic animals (Dove 2002).

Plant pharming is being used by several companies and universities to develop vaccines and therapeutics. Although transgenic plants also offer significant cost and scalability advantages over cell culture systems, plant-based transgenics are generally less mature than animal-based systems; further, plant-based companies have experienced limited progress in transitioning products to human clinical trials.

The major hurdles to the adoption of plant-based transgenics are safety concerns, both health and environmental. Health concerns include the probability of immune responses caused by the attachment of the therapeutic proteins with a complex carbohydrate. Environmental concerns are mostly focused on the spread of transgenics in nature.

Some companies are confining their activities to greenhouses, but approximately twenty transgenic plant firms are currently planning or conducting open-air field trials of their plants. Although these companies are proceeding with the utmost caution regarding environmental issues, there are significant concerns associated with the potential impact on animals, insects, soil microorganisms, and neighboring crops. However, public concern over using transgenic plants for drug production has generally been less of an issue compared to other genetically engineered plants used for food, primarily due to the fact that drug-producing plants are viewed more as prescription drugs than as foods (Van Brunt 2002).

Transgenic plants have shown the potential for producing large-scale genetically engineered proteins. A key technical issue, though, is that these proteins are not glycosylated and have not been processed through the normal pathways that a mammalian protein would be. As a result, transgenic plants will likely be a viable alternative for some proteins that are less complex and do not require the folding and glycosylation that a mammalian system provides. In fact, David Cerny, director of business development for Epicyte Pharmaceutical, Inc., noted that only two of the monoclonal antibodies currently on the market require mammalian glycosylation to elicit complementary responses. However, if postproduction methods of folding, glycosylating, or otherwise altering proteins become available, then plant-based systems could become leading candidates for large-scale protein production. This method ensures there is no chance that animals will be mistreated or that animal disease will be passed through the protein (Cerny 2002).

COMMERCIAL DEVELOPMENT OF TRANSGENIC PROTEINS

Given the imminent shortages in biologics manufacturing capacity, there is a growing need for more efficient production systems. Several emerging biotechnology companies have focused on transgenic protein production over the past several years and have achieved varying levels of success. But despite the potential of transgenic protein production, interest in transgenic-based companies on Wall Street and in the investment community has continued to diminish. The primary reasons for the declining interest in transgenics companies have been the high degree of uncertainty and the long time frame associated with attaining FDA approval of a transgenically produced protein (Rosenblum 2002).

Currently, the most advanced competitors in the transgenics sector are GTC Biotherapeutics, PPL Therapeutics, and Pharming Group N.V. (Figure 7.2). The companies have cross-licensed rights to several patents. For example, GTC, PPL, and Pharming have cross-licensed rights to the promoter sequences enabling the production of human proteins in milk in various animals.

GTC Biotherapeutics, Framingham, Massachusetts (formerly Genzyme Transgenics): GTC specializes in transgenic technology in animals and has established a leadership position in the field. The company developed a broad roster of corporate clients, including Bristol-Myers Squibb, Centocor (subsidiary of Johnson and Johnson), Eli Lilly, BASF Knoll, and Fresenius. The company established a strong intellectual property position in antibody production, an area expected to experience significant growth in the future. To date, GTC has produced 65 proteins transgenically, including approximately 16 antibodies. Of the 65 proteins produced by GTC in transgenic animals, 45 of them have been expressed at levels greater than 1 gram per liter. The company is also expecting completion of the first pivotal study with a transgenic protein, AT-III, in the near future. GTC works primarily with transgenic goats because of their relatively high milk yield and the short time required for maturation. To help address the purification requirement issue, GTC established a strategic alliance with Lonza Ltd. to develop purification and production methods for proteins. Over 70 of the 369 proteins currently in clinical trials are monoclonal antibodies. GTC is, then, well positioned to achieve market penetration with its technology.

PPL Therapeutics, PLC, Edinburgh, UK: PPL Therapeutics utilizes nuclear transfer technology to create transgenic sheep for the production of transgenic proteins. The company garnered worldwide acclaim from its work with the Roslin Institute in cloning the world's first sheep, named Dolly. PPL discovers, develops, and produces recombinant human proteins for therapeutic and nutritional use. Some of the company's products are novel and already approved for human use, albeit currently produced from human plasma. PPL's strategy is to market its products through collaborations with major pharmaceutical companies. PPL is also pursuing production using cows as a result of the European Patent Office's declaration that pharming's key European patent on the use of cows for producing protein in milk was null and void because it lacked any inventive step beyond what was already known.

Pharming Group N.V., Al Leiden, the Netherlands: Pharming has four business units: orphan products, specialty products, plasma products, and transgenic technologies. The company is a Dutch biopharmaceutical firm that is developing therapeutics for treatment of various genetic disorders, infectious diseases, tissue and bone damage, cardiovascular diseases, surgery, and trauma. Pharming has been issued thirty-one patents related to transgenic technology and products derived from the milk of transgenic animals. Interestingly, Pharming has offered a nonexclusive licensing agreement to GTC related to the production of human serum albumin in cattle.

There is also competition from other emerging companies focused on transgenic protein production using mammals, chickens, and plants, including Nexia Biotechnologies, Inc. (mammals), Gala Design (mammals), Epicyte (plants), Avigenics (chickens), Advanced Cell Therapeutics, ProdiGene, and Monsanto.

Some companies are using chickens to compete with other transgenic mammalian systems on cost and differentiate themselves on speed to market and scalability. It requires significantly less time to generate a chicken flock than a herd of goats or cows.

Figure 7.2   Most Prominent Pharming Companies

ADVANTAGES OF TRANSGENIC PRODUCTION

ECONOMIC ADVANTAGES

As discussed in the previous chapter, drugmakers find it difficult to justify biologics due, in part, to the high cost of manufacturing. However, the ability to produce and recover recombinant proteins from the milk or urine of transgenic animals has offered the potential for a substantially more cost-effective and faster production method.

Table 7.2 illustrates a comparison of the approximate manufacturing costs, including capital and operating costs, of biopharmaceuticals produced using bacterial fermentation, mammalian cell culture, and the milk of transgenic animals. At larger production volumes, the manufacturing costs of biologics using bacterial fermentation and transgenic animal milk are roughly equivalent, but many complex proteins cannot be expressed in active form by bacterial fermentation.

The economic factors associated with transgenic production can potentially generate a lower overall cost of goods sold for the biologic being produced. High protein expression levels in transgenic animals and efficiency in purification typically result in making the cost of transgenically produced protein significantly lower than that

Table 7.2
Manufacturing Costs of Biologics (US$/gram of protein)*

	Scale (kilograms)			
	0–10	*10–100*	*100–1,000*	*>1,000*
Bacterial fermentation†	$22,000	—	$40–65	$32–55
Mammalian cell culture	10,500	$300–3,000	—	1,000–5,000
Transgenic animal (milk)	300–500	100–300	30–100	<40

Source: Nexia Biotechnologies Prospectus 2000. (Prospectus can be requested from Nexia Biotechnologies, Vaudrevil-Dorion, Quebec, by calling 450-424-3067 or email lr@nexiabiotech.com.)

* These costs are for bulk active purified proteins but do not include the fill and finishing costs associated with the production of biopharmaceuticals.

† Many complex proteins cannot be expressed in active form by bacterial fermentation.

— = no cost

of product manufactured in cell cultures. Genzyme Transgenics and Centocor (Johnson and Johnson) are working together on the production of Centocor's drug Remicade in transgenic goats. According to Dr. Frederick Bader, Centocor's vice president of process science and engineering, the capital spending for transgenic production of their product was $80 million less than for cell culture facilities. Furthermore, Bader stated that the estimated cost for transgenic production would be $100 per gram of protein, whereas cell culture production would cost approximately $333 per gram. These operational cost savings were expected to result in a payback of the fixed cost over three years for annual production levels in the 40 to 50 kilogram range (Burrill and Company 2001: 139–140).

A significant portion of the cost of manufacturing systems lies in the up-front capital investment required for the manufacturing facility, a cost that is much lower with transgenic animals. Developing a transgenic herd and establishing appropriate dairy facilities operated within accepted agricultural practices can usually be accomplished with substantially less capital investment than constructing a cell culture bioreactor facility. Although the cost of constructing cell culture facilities has declined modestly over time, recent plant construction activity in the industry indicates that the level of capital required for commercial-scale facilities is approximately $2 million per 1,000 liters of capacity. This figure implies that $5 billion to $6 billion of capital will be required to meet the pharmaceutical industry's expected production requirements (Andersson and Mynahan 2001: 2).

As an example, Genentech invested $250 million to construct and develop a mammalian cell culture manufacturing facility with a capacity in excess of 1,000 kilograms of antibody material in Vacaville, California. According to Roger Lias of Covance Biotechnology Services, there may be up to a hundred new antibodies approved by 2010, which would require approximately sixteen times the capacity of Genentech's Vacaville facility (Burrill and Company 2001: 140).

Costs associated with these cell culture systems can range from around $200 to $1,000 per gram, with capital expenditures of up to $50 million. In comparison, it costs approximately $20,000 to deliver a milking goat to the herd, and the total capital expenditure requirements to build facilities for the herd are roughly between $5 million and $7 million (Gavin 2001).

One of the primary economic benefits associated with transgenic animals involves the scalability of the technology, given the significantly reduced amount of time and capital required to expand a production herd as compared to constructing a cell culture plant. At a cost of approximately $5 million, a transgenic herd can be developed to produce approximately 100 kilograms of partially purified protein. An equal amount of protein would require as much as $20 million in capital spending for a typical Chinese hamster's ovary (CHO) cell-based facility. Beyond start-up costs, the leverage in increasing scale is greater for transgenic animals. For cell cultures, a fivefold increase in production would require over three times the amount of initial capital and would yield a 60 percent improvement in margin. In contrast, the same increase in capacity with a transgenic herd requires only two times the amount of initial capital and generates the same improvement in margin (Tanner, Stading, and Jacobson 2000).

TECHNICAL ADVANTAGES

The limitations of traditional protein production technologies have led to the increasing application of transgenic technology in manufacturing complex proteins in large quantities. Transgenic animals are particularly suitable for producing complex glycosylated proteins, such as monoclonal antibody (mAb) and immunoglobin (Ig) fusion molecules, in which two pairs of distinct protein chains must correctly assemble in vivo for proper function. Although effective for large-scale production, bacteria are unable to add sugar molecules to proteins (glycosylation), a modification that is important with respect to a protein's function in the body (e.g., half-life and recognition of binding proteins). Mammalian cell culture technologies were subsequently developed with the capability to glycosylate proteins, but they are generally less efficient in production than transgenic animals and are not as economically feasible under certain circumstances given their lack of scalability.

Recent advancements in transgenic technologies have led to the ability to scale herds at a linear rate. By generating a founder animal with the transgenic protein encoded in the genetic makeup of the animal, the protein can be passed on to future generation by simple breeding. This allows a scale-up of manufacturing in an easy and reliable manner, with only simple tests required to ensure proper protein production.

MARKET ADVANTAGES

The worldwide market for protein manufacturing represents a substantial and growing portion of the value chain for pharmaceutical companies and presents an attractive opportunity for transgenic-based manufacturers to expand their businesses. Although transgenic production comprises a small portion of the current market, the impending shortage in traditional manufacturing capacity will likely compel pharmaceutical companies to reassess their near-term production strategies and provide a substantial near-term opportunity for transgenic-based companies to gain market penetration. In addition to the rapidly growing contract manufacturing market, in-house manufacturing is expected to continue to grow rapidly and contribute to a total protein manufacturing market of over $6.4 billion by 2005 (Clark et al. 2001).

VALIDATION ADVANTAGES

Several companies focused on developing commercial therapeutics using transgenic technology have been validated in the marketplace through partnerships or acquisitions with established pharmaceutical companies and through advancements in clinical trials. For instance, GTC entered into several partnerships to develop biopharmaceuticals transgenically with established pharmaceutical and biotechnology companies, including Bristol-Myers Squibb, Elan, Centocor, Abgenix, and Alexion. In addition, Pharming Group entered into partnerships with Baxter and Genzyme General, and PPL Therapeutics has partnered with Bayer.

PRICING ADVANTAGES

As pharmaceutical companies continue to experience regulatory and social pricing pressure for their products, the use of transgenic technology to lower their drug manufacturing costs could provide additional pricing flexibility and expand the target markets for their products. For example, the proliferation of transgenic technology potentially has significant implications for providing therapeutics to developing countries. A program being conducted by GCT focuses on developing malaria antigens for vaccines. These antigens, which are difficult to manufacture using cell culture technology, were recently produced in mice and protected monkeys from malaria. The company predicts that if goats can also produce the vaccine, then only three of the animals could supply enough vaccine to immunize 20 million African children (Sedlak 2002: 21).

Lower-cost transgenic production also has the potential to offer unique opportunities for profitably producing generic forms of existing biologics that are coming off patent. Although firms will have to overcome bioequivalence issues to manufacture a generic product, clinical trials for generic versions of a drug are typically less onerous than the initial trials because the doses and side effects are already known. Nexia Biotechnologies, for example, is planning to supply a low-cost generic version of human tPA (tissue plasminogen activator). Human tPA is off patent in Canada currently and was not covered in the United Kingdom in 2002 and in the European Union in 2003, nor will it be covered in the United States in 2005. Nexia currently has a small herd that can produce 20 kilograms of tPA annually, which is a large enough supply of the drug to meet demand worldwide (Sedlak 2002: 43).

POTENTIAL RISKS ASSOCIATED WITH
TRANSGENIC PRODUCTION

We identified eight important market and nonmarket (regulatory, political, and public) risks associated with pharming.

REGULATIONS

The regulatory environment for the protein manufacturing industry using transgenic animals and plants is still somewhat unclear. Regulation will probably feature a combination of FDA rules that govern cell culture manufacturing, moral animal treatment, and protein purification testing. Both the FDA and the European regulatory agencies have issued documents with guidelines for the safe production of proteins using pharming practices. The standard FDA process involving preclinical laboratory and animal testing and submission of an investigational new drug (IND) application or a biologics license application (BLA) will still be required as a basis for drug approval. For animal-based transgenics companies, the FDA is expected to be the primary regulatory body; plant-based companies will also be subject to additional U.S. Department of Agriculture (USDA) and Environmental Protection Agency (EPA) regulations.

GCT is the only company to have submitted to the FDA a protein produced in the milk of a transgenic animal, and a number of similar products are currently in advanced human clinical trials. Although GCT's product was not approved, it is believed that it was denied for reasons other than transgenic protein production.

Protein-based products that have gone off patent or improved formulations of a drug can achieve approval by the FDA as well, likely through an expedited process. Companies can rely on the published studies of others for safety and efficacy results in this case. There will be a number of biological products coming off patent in the coming years, and pharming companies may be able to use this process as a way to expedite their own efforts to get a product to market.

CONTAMINATION

Contamination is a primary concern with transgenic animals, as there is the potential risk of pathogens passing into the milk and contaminating the protein. As a result, paying attention to the stock of animals will play a key role in assuring a disease-free herd. Animals from countries that are free of livestock diseases, such as New Zealand or Australia, will have significant advantages with the regulatory agencies. Keeping multiple herds will also be critical to this strategy in case there is a disease in one area that requires the animals be destroyed. The additional cost involved can be viewed as insurance but also will allow for a further production of protein in two separate facilities.

With regard to transgenic plants, ecological contamination is also a major issue. Companies pursuing transgenic plant production will have to follow strict containment procedures to prevent "gene flow"—the transfer of transgenic genes to neighboring crops through cross-pollination. Similarly, companies will likely be forced to ensure the safety of their transgenic crops in terms of the potential interaction of the genetically engineered drugs, vaccines, and other substances in the plants with foraging animals, seed-eating birds, and insects in the environment.

REIMBURSEMENT

How transgenically produced therapeutics will be reimbursed is still unknown. Efforts to achieve cost containment may lead to predetermined low prices and discounts. Essentially, if the proteins can be produced at a low cost, why should those savings not be passed along to the consumer, eliminating the additional profit margin of the companies pursuing pharming?

The great advantage of this technology lies in the potential cost savings. However, if this advantage cannot be realized due to decreased government and insurance reimbursement, then the advantage is lost. In the current environment of increasing expenditures on pharmaceutical products, this may be an area where low reimbursement takes place. Once the technology is in place to reduce the cost of protein production, government agencies and insurance companies will have an argument to reduce reimbursement rates.

SOCIAL CONCERNS

Public acceptance is a challenging issue for transgenics. The degree of market acceptance will depend on a number of factors, including cost of the product, safety and efficacy of the proteins, regulatory approval, and approval by nongovernmental organizations (NGOs). There is no assurance that products developed using recombinant proteins will ever gain market acceptance.

Various animal rights groups and other organizations and individuals have attempted to stop animal testing and genetic engineering activities by boycotting or pressing for legislation and regulation in these areas and engaging in public demonstrations and media campaigns. Unfortunately, transgenic technology does carry with it the capacity to cause suffering to the animals involved. The manipulations of genes themselves can harm the animals. Insertion of a gene has caused problems in the past due to overexpression in inappropriate tissues. Problems have also arisen from disruptions of the normal gene functions, leading to undesirable effects. The technology today is somewhat inefficient, and there appear to be ways to reduce the cost of the process. With a current success rate of between 5 to 25 percent, it takes a number of attempts to produce a successful transgenic animal with the gene inserted in the appropriate place.

PARTNERSHIP VULNERABILITY

Dependence on strong alliances and licensing deals to fund development is not unique to pharming companies. However, because the technology is so new, these deals are more critical than deals with traditional drug discovery biotechs. Many pharmaceutical companies exploring transgenic production tentatively depend on collaborative and licensing arrangements with the existing biotechs developing transgenic animals and plants. If these companies fail, for either financial or technical reasons, pharmaceutical companies will immediately regard that failure as an omen that pharming does not work and will pull out of the alliances.

INTELLECTUAL PROPERTY

Manufacturing processes usually have weak intellectual property positions. However, pharming companies seem to have developed strong intellectual property platforms. But their status is vulnerable. Transgenics-based companies essentially have patents around engineering processes for manufacturing proteins, and engineering patents are notoriously easy to reverse engineer. Some of the companies have patents for promoter sequences for certain animals, but there are a large number of animals available for milk production.

COMPETITION

Competition is fierce in this arena. The pace of technological advancement is likely to be rapid, making it essential for companies to partner with the leading providers of transgenic technology. There are a few main players in the transgenic animal field, and a number of companies are working on alternative technologies. There are also several large contract manufacturers of proteins that must be considered competitors to transgenic production. In addition, transgenic production may prove to be less attractive if there are technical improvements in mammalian cell culture methods, such as more efficient bioreactors and enhancements in cell lines, media, and purification technologies.

DISRUPTIONS

The emergence of a disruptive technology that would provide a safer or more cost-effective production of biologics is always a possibility and, by definition, impossible to predict.

GOVERNMENTAL REGULATION

In 1986, the Office of Science and Technology Policy published the *Coordinated Framework for the Regulation of Biotechnology,* which considered existing regulations applicable to biotechnology and proposed how the FDA, the USDA, and the EPA would cooperate in the review of new biotechnology. This framework is still in existence today and establishes the basis for regulation of new animal and plant varieties produced by genetic engineering techniques (American Medical Association 2000) (Table 7.3).

The most applicable regulation for protein production in transgenic animals comes from the FDA's oversight of human drugs and medical devices. However, the USDA oversees regulation regarding the production and approval of biologic agents such as those produced in sterile manufacturing facilities and existing agricultural regulation regarding clean, safe facilities and practices for maintaining herds for human food products. Some industry practitioners believe that completely new statutory frameworks designed specifically for transgenic products are not necessary because the FDA has the legal authority to regulate such products.

Definitive guidelines for transgenic plants have yet to be issued by the USDA, but the FDA has produced several drafts of a paper entitled "Points to Consider in the Manufacture and Testing of Therapeutic Products for Human Use Derived from Transgenic Animals." We will discuss a summary of this "Points to Consider" document because it is the most relevant regulation in this field and because it also highlights several areas where executives must make certain cost-benefit decisions that we feel will be challenging, to say the least.

Table 7.3
Regulatory Agencies for Biotechnology Products

Agency	Products Regulated
U.S. Department of Agriculture	Plant pests, plants, veterinary biologics
Environmental Protection Agency	Microbial/plant pesticides, new uses of existing pesticides, novel microorganisms
Food and Drug Administration	Food, feed, food additives, veterinary drugs, human drugs, and medical devices

Source: http://www.usda.gov/agencies/biotech/faq.html.

Current FDA definitions classify a transgenic animal as an animal "which is altered by the introduction of recombinant DNA through human intervention" ("Points to Consider" 1995). This definition includes both animals with heritable germ-line DNA alterations and those with somatic nonheritable alterations. The thrust of the "Points to Consider" document is to guide companies in the types and amount of information that is gathered regarding the creation of transgenic animals and production herds, the care and maintenance of those animals, and the quality and purity of the final protein product. Though very few regulations are in place that specifically focus on transgenic animals, the "Points to Consider" document provides insight into the information necessary to successfully complete the FDA approval process for new human therapeutics.

The sheer volume of information that must be collected and maintained is an area on which management must focus. Management is presented with unique challenges because the required information spans many different disciplines, ranging from DNA sequencing to transgene insertion to monitoring and maintaining a healthy livestock herd. In addition, much of this information must be collected before a product is even considered for clinical trials. Some of the issues addressed in the document that present significant managerial challenges are discussed in the following sections.

GENERATION AND CHARACTERIZATION OF THE TRANSGENE CONSTRUCT

Information must be collected long before a transgenic animal is produced. This presents a challenge to management to ensure that proper documentation procedures are followed from the time an idea is initially considered. Companies attempting to create transgenic animals should keep detailed information regarding the original gene, expression patterns, the transgene construct (including details on quality control of the construct assembly), cloning, and purification, as well as details relating to the original vector.

CREATION AND CHARACTERIZATION OF THE TRANSGENIC FOUNDER ANIMAL

Managers in the biotechnology or pharmaceutical industries must track information relating to the genealogical and health history of donor animals, an area to which they may not have been previously exposed. The FDA tries to ensure that specific diseases, such as spongiform encephalopathy, are not introduced into production animals, thereby minimizing the risk of these diseases being passed on to humans. Based on the "Points to Consider" document, there is a host of other information that must be tracked and challenges that must be overcome regarding the founder animal, including the following:

- The methods of introduction of the DNA and the tests for the presence of the transgene in the founder
- The identification of animals that have taken up exogenous DNA but are not producing transgenic products
- The identification of seasonal, age-related, or other variations in product
- The stability of the transgene and transgene expression through several rounds of breeding

ESTABLISHMENT OF A RELIABLE AND CONTINUOUS SOURCE OF TRANSGENIC ANIMALS (FOUNDER STRAIN)

Management must address the issue of ensuring that production of the desired product remains available for an extended period of time. Unlike cell cultures, which can be stored indefinitely, animals have a finite life span. Further complications involve the different interactions of founder animal genes with genes of different breeding partners. For this reason, the FDA proposed a solution similar to that used in biological production cell lines. A master transgenic bank consisting of a limited number of highly characterized animals (or sperm and embryos) derived from a particular founder that can reliably yield product-producing offspring may allow companies to maintain a stable supply of genetic material from which to ensure continued protein production.

GENERATION AND SELECTION OF THE PRODUCTION HERDS

The FDA guidance is vague in this regard, so managers face the challenge of establishing for themselves procedures and testing standards that animals must pass before being added to a production herd. Minimum guidance includes animal history and genealogy (traceable to a founder animal), breeding techniques, and minimum production quantities. We believe the FDA will eventually require reintroduction procedures and standards of previously removed animals to be more stringent than initial introduction requirements for production herd animals. An additional burden of proof is placed on reintroduction because a removal event will most likely be caused by an illness. For reintroduction, strong evidence must be shown to prove that the illness has been eliminated from the animal and that the product has no lingering effects. In contrast, an animal with a good health history does not have to overcome this added burden in order to gain acceptance into the herd.

MAINTENANCE OF TRANSGENIC ANIMALS

There are specific guidelines for the maintenance and disposal of animals and their by-products, as well as guidelines covering their housing facilities. Practices at a transgenic production facility must be documented and approved according to the Animal Welfare Act. We believe that companies intending to use transgenic animals should find and hire managers already experienced in these sound livestock farming practices. The cost of a mistake, in the form of lost revenue caused by a facility shutdown, dwarfs the cost of finding and hiring an experienced manager who can greatly improve the odds of avoiding such a mistake. In addition, we believe there is a large pool of talent from which to draw due to the myriad traditional farming operations in the United States. The "Points to Consider" document also provides limited guidance on specific requirements for transgenic production facilities. Such guidance includes the documentation of complete histories of all production animals, from birth to death, and the effective monitoring and control procedures that evaluate the expression of the product.

PURIFICATION AND CHARACTERIZATION OF THE TRANSGENIC PRODUCT

This is an area in which managers are faced with extremely vague guidance from the FDA. Although the "Points to Consider" document acknowledges that it is impractical to maintain sterile conditions when collecting biologics from a large number of animals, it suggests that facilities should still be "as clean as possible." This vagueness is particularly challenging as managers see decreasing returns on incremental investments made to improve the cleanliness of production facilities. A lack of experience with many proposed host animals will cause the FDA to consider submissions on a case-by-case basis, thus increasing the ambiguity under which management must operate.

LOT-TO-LOT VARIATION

Lot-to-lot variation is expected to be higher in proteins produced from transgenic animals compared to those produced in continuous cell cultures and other production technologies. Management will face challenges in proving that each lot produces a safe and consistent product. In addition to controlling for lot-to-lot variation, products derived from transgenic animals have the potential to contain a variety of human pathogens that must be screened. The greatest sensitivity of these tests is realized when performed at the single animal level, but this form of testing happens to be the most expensive. Management must determine at what stage of production the greatest cost-benefit balance is realized. In addition, the FDA recognizes that products may contain not only species-specific pathogens but also medications and other undesired particles. As a result, management will need to adopt appropriate testing methods to screen for many of these impurities as well.

The "Points to Consider" document does not provide clear rules on a great many issues regarding the production of proteins in transgenic animals. The area where current regulation is very clear happens to be an area in which biotechnology and pharmaceutical managers have little experience, mainly involving the proper maintenance of herds and herd facilities. Future regulation has the potential to affect the cost structure of the pharming industry. For instance, if future regulation requires pathogen screening at the individual animal level as opposed to the pooled production level, the cost of such a requirement could make the overall cost of transgenic production higher than that of competing production technologies. Companies seeking to enter the pharming industry must pay particular attention to proposed regulation and take actions to ensure that reasonable regulation is enacted.

DECISION FRAMEWORK

As the demand for protein-based drugs is expected to substantially outpace manufacturing capacity in the near term, pharmaceutical company executives will have to make important decisions regarding how to build or buy new manufacturing capacity utilizing either traditional cell culture technologies or transgenic organisms. In addition to satisfying their own manufacturing requirements, it will also be critical for pharmaceutical companies to possess flexible production capabilities in order to maintain their attractiveness as potential partners for other drug developers. Many of the late-stage biologics in the pipeline are owned by emerging biotechnology companies that do not have the ability to produce and market their products internally. As a result, pharmaceutical companies that can offer manufacturing capacity in addition to sales and marketing capabilities will be able to approach partnering from a position of strength (Andersson and Mynahan 2001: 3).

Given that transgenic protein production is a relatively new manufacturing process, executives must also consider the potential timing of making investments in transgenic technology. Under certain circumstances, it may be prudent to wait several years until a transgenically produced biologic has been approved by the government and the technology has matured. In other cases, near-term supply needs may necessitate immediate investments in transgenic production, which will involve a greater amount of technological and regulatory risk.

One of the primary reasons pharmaceutical companies are concerned about investing in transgenic protein production methods is the fact that there are currently other acceptable methods for manufacturing proteins. The potential advantages of transgenic production include cost savings, enhanced scalability, and the ability to produce complex proteins that no other current manufacturing system can make. However, if other technologies advance or emerge that can eliminate any of these advantages, then transgenic-based strategies could be at a great disadvantage. For example, cell culture systems have the potential to become more efficient, making them an even more attractive option relative to transgenic production. U.S. Bancorp Piper Jaffray has estimated there will be more than a threefold improvement in cell culture production rates due to improved cell lines, more runs per year, and higher yields. Such improvements in traditional manufacturing methods could reduce the potential of transgenics to a much smaller, niche market (Ginsberg, Bhatia, and McMinn 2002: 15).

Given the proven track record of cell culture technologies and the potential of alternative production systems, the benefits of transgenic protein production will have to significantly outweigh the development and regulatory risks associated with the technology in order to achieve widespread adoption among pharmaceutical companies. This fact is reflected in the recent decision by IDEC Pharmaceuticals (now Biogen-IDEC) to invest over $400 million in a new plant to produce its new antibody therapeutics. William Rastetter, past chief executive officer (CEO) of IDEC Pharmaceuticals, noted that as his company evaluated transgenic production for commercial purposes, the actual costs associated with transgenics were significantly higher than expected due to substantial purification issues. Rastetter explained that the existing difficulties associated with passing equivalence tests for different cell culture facilities manufacturing the same product would likely be magnitudes greater for drugs produced transgenically, given potential variations in individual animals. In addition, he emphasized the expected improvements in the productivity of cell culture systems and stated he believed that the regulatory risk associated with transgenically produced drugs continues to significantly outweigh any potential cost benefits that pharming might offer (Rastetter 2002).

A decision framework for pharmaceutical executives to consider when facing a production decision is shown in Figure 7.3. Several key issues come into play, including the complexity of the protein (e.g., glycosylation, folding), the volume requirements, and the reliability of the system. Based on the substantial amount of uncertainty surrounding the regulatory approval of transgenically produced biopharmaceuticals, as well as potentially improving economics and production from mammalian cell culture facilities, we believe that transgenic production is currently an attractive option only for those complex proteins that require extremely large volumes of capacity or that cannot be produced by traditional manufacturing systems.

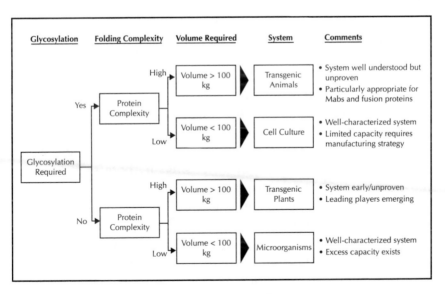

Figure 7.3   Manufacturing Decision Framework

Assuming conditions warrant the use of transgenic-based systems for manufacturing, we believe that management should take the following courses of action to help ensure favorable regulatory treatment and public acceptance of transgenically produced therapeutics.

PROACTIVELY MANAGE REGULATORY GUIDELINES

The growth of this industry has produced an immediate need for clear guidance from the FDA. The original "Points to Consider" document was drafted several years ago, at a time when this production technology was less well developed. An immediate dialogue with the FDA is necessary to provide up-to-date guidance. Given the vagueness of existing regulations, companies pursuing production in transgenic animals must focus on influencing regulations as they are developed. Companies should adopt the following strategies when addressing regulatory frameworks:

- Pursue lobbying efforts to focus on providing clear guidelines for transgenic production
- Encourage the FDA to adopt "middle-of-the-road" testing procedures that allow companies to realize cost benefits but require appropriate testing levels to discourage less reputable companies from entering this space
- Immediately pursue an active dialogue with the FDA and USDA concerning future regulation

A clear set of guidelines is needed for managers to know exactly what requirements must be met. Even if those guidelines are more stringent than what some companies desire, they are better than unclear guidelines because they will help managers make effective resource allocation decisions. The current guidance from the FDA does not allow managers and executives to weigh investment decisions against probabilities of negative FDA reactions.

Initial industry reactions to regulation may push regulators toward allowing the least stringent testing. We believe the industry should take a longer-term view and push for regulations that permit companies to realize profit potential while providing a framework that allows product safety to be verified. This type of regulation will provide legitimacy for the industry by preventing firms without a long-term focus from entering. In addition, this level of regulation will help address public concerns over the use of animals to manufacture therapeutic proteins.

PROACTIVELY MANAGE PUBLIC ACCEPTANCE OF TRANSGENIC TECHNOLOGY

We also anticipate that many companies and management teams will be concerned about the public acceptance of drugs manufactured from transgenic organisms. Some groups will likely argue that if there are methods of production available that do not harm animals or impact the environment in any way, those should be the production methods of choice. We believe that management will be faced with issues raised by animal rights groups and other NGOs that have taken public positions against producing genetically modified animals and have argued that such animals are being abused for the good of humans.

It is our belief that the best way to diffuse these arguments is to produce a human therapeutic protein product that has yet to be produced using any other method. If the industry develops a truly novel compound that cannot be made through cell culture methods because the protein is too complex, then much of the criticism of the NGOs will be deflected. This would allow the transgenic protein industry to obtain a foothold and to prove itself while helping save or improve lives. Once therapeutic protein technology has proven itself to be safe, inexpensive, and cruelty-free to animals, the public may be willing to accept more widespread use of the technology in manufacturing biopharmaceuticals.

However, the industry must balance positive public acceptance with appropriate business needs. An alternative approach would entail entering this market with either a nontherapeutic protein or with a protein that has been used therapeutically for many years and no longer has a patent. This strategy has been pursued by Nexia

Biotechnologies, a company that uses transgenic goats to produce both spider silk and biopharmaceuticals. In a *Forbes* article, Jeffrey Turner, CEO of Nexia, stated that manufacturing drugs with goats is such a pie-in-the-sky dream that Nexia is allocating only 20 percent of its research and development budget into creating such drugs. A nontherapeutic protein would also allow a company to make mistakes while learning effective processes and controls without the threat that a small mistake will lead to an entire operational shutdown by the FDA during a drug approval process (Herper 2002).

A generic therapeutic protein would allow the company to go through trials designed to prove only the safety of the protein, which would be less expensive than a full FDA trial. Once managers gained experience in the breeding and production of proteins, they would be better equipped to manage processes in conformity with FDA guidance for therapeutic protein production. However, this approach is directly contrary to what we feel is the best approach for this industry from a public acceptance standpoint. To minimize criticism from groups such as People for the Ethical Treatment of Animals (PETA) and to gain the most favorable public acceptance, we believe that pursuing the production of a therapeutic protein that has not yet been produced and cannot be produced by other methods is the best alternative. Although the generic market may likely prove to be the biggest market for transgenically derived proteins as well as a potentially huge opportunity for these companies, there will always be concern for animal and human safety until this method of production has been proven to be safe for both. Demonstrating the safety to the public can be done, but we feel the public will be most open to accepting the risk entailed in this technology if the product cannot be produced using an alternative method.

We also anticipate that management will seek to exploit the number of protein-based products coming off patent in the years ahead. This opportunity for competitive products will likely represent a large potential market for a low-cost manufacturer of proteins, and it would be an outstanding way to put transgenic protein production technologies to the test. These products would not require long clinical trials to prove their efficacy, having only to prove bioequivalence and safety. This would allow the transgenic companies to compete on the basis of low-cost production, which is their primary advantage.

REFERENCES

American Medical Association. 2000. "Genetically Modified Crops and Foods," December, 1–25. Available at http://www.ama.assn.org/ama/pub/article/2036-4030.html. Accessed June 2003.

Andersson, R., and R. Mynahan. 2001. "The Protein Production Challenge." In Vivo (Winhover Information, Inc., newsletter), May.

"Approved Biotechnology Drugs." 2004. Available at http://www.bio.org/speeches/pubs/er/approveddrugs.asp. Accessed June 2004.

Burrill and Company. 2001. Biotech 2001: Annual Biotechnology Industry Report. San Francisco: Burrill and Company.

Cerny, D. 2002. Interview on March 6.

Clark, P., G. Haire, R. Tucker, and E. Illidge. 2001. "Lonza Group AG." UBS Warburg Report, February 20, 5–7.

Dove, A. 2002. "Uncorking the Biomanufacturing Bottleneck." Nature/Biotechnology 20: 777–779.

Fox, S., R. Khoury, and L. Sopchak. 2001. "A Time to Build Capacity." Contract Pharma. September, 1–6. Available at http://www.contractpharma.com/sept011.htm.

"Frequently Asked Questions—USDA and Biotechnology." N.d. Available at http://www.usda.gov/agencies/biotech/faq.html. Accessed June 2004.

Gavin, W. G. 2001. "The Future of Transgenics." Regulatory Affairs Focus publication, May.

Ginsberg, P. L., S. Bhatia, and R. McMinn. 2002. "The Road Ahead for Biologics Manufacturing." Piper Jaffrey Report, January.

Herper, M. 2002. "Milking Genetically Modified Cows." Forbes.com. April 1, 1.

Nexia Biotechnologies Prospectus. 2000. (Contact Nexia Biotechnologies at [450] 424-3067.)

"Points to Consider in the Manufacture and Testing of Therapeutic Products for Humans Derived from Transgenic Animals." 1995. U.S. Food and Drug Administration Center for Biologics Evaluation and Research. Available at http://mbcr.bcm.tmc.edu/BEP/ERMB/ptc_tga.html# definition.

Rastetter, W. 2002. Interview on March 18.

Rosenblum, S. (research analyst with Lehman Brothers, Inc.). 2002. Interview on March 13.

Sedlak, B. J. 2002. "Protein Manufacturing in Transgenic Animals." *Genetic Engineering News*, January.

Tanner, W., T. Stading, and E. Jacobson. 2000. "Genzyme Transgenics." SG Cowen Report, June.

Van Brunt, J. 2002. "Molecular Farming's Factories." *Signals Magazine.* February 19, 2.

Chapter 8

THE ROLE OF
MARKETING

Dario Benavides, Lanier Coles,
Gary Dulaney, Vanessa Rath,
and Carla Yarger

Once a new drug application (NDA) or biologic license application (BLA) is submitted to the U.S. Food and Drug Administration (FDA), the role that marketing plays in biotechnology and pharmaceutical companies becomes apparent. The traditional marketing functions involving product placement, price, and promotion become central to the success of the new product's launch. It typically takes twelve to eighteen months of planning to effectively prepare for a product launch (Garret 2002). Early collaboration between the research and development (R&D) and marketing arms of the company, via market planning and development, is crucial as well, but this aspect is often overlooked by small biotechnology firms. Conversely, pharmaceutical companies have a better understanding of the importance of this relationship.

By evaluating biotechnology and pharmaceutical firms and their successes and failures in new product launches, we can begin to understand the impact of marketing in the early development of new products. In this chapter, we will first present a series of general observations on the current status of the marketing role in biotechnology. Then we will evaluate the marketing function for three biotechnology/pharmaceutical products (Viagra, Enbrel, and

Natrecor) and examine how that function was incorporated or neglected in the development of these drugs. From these development cycle profiles, we will formulate our hypothesis that sophisticated early-stage marketing enhances the value of the products being developed. Further, we will postulate that both biotechnology and pharmaceutical firms stand to create more productive alliances if they attain a solid understanding of their market opportunities before entering such partnerships. Finally, we will make several recommendations for optimizing marketing efforts in the development of biotechnology and pharmaceutical products. These observations and recommendations are based on a series of interviews and an analysis of a panel of thirty biotechnology companies (Garrett 2003).

HOW MARKETING HAS EVOLVED IN BIOTECHNOLOGY

The biotechnology industry has been undergoing a significant transformation since the late 1980s. The traditional business model, which included projected earnings ten to fifteen years down the road, is no longer viable. In the former era, biotechnology companies were predominately purely research-based companies, with little focus on marketing and commercialization. In contrast, today's biotechnology firms are adopting a more pragmatic model to appease their shareholders. They have come to realize that a successful product launch is strongly correlated with a clear understanding of the marketing and commercialization issues associated with the product and its potential competitive arena. Because this is typically not a core competency of biotechnology firms, many companies are forming alliances with pharmaceutical firms, traditionally recognized as having strong marketing and sales infrastructures. In 2002, there were thirty-one deals between pharmaceutical and biotechnology firms valued at over $200 billion; although this represents a decrease from the absolute number of deals in 2001, the overall total value of the deals increased (Burrill 2003). This new model seeks to introduce earlier revenue streams to cash-poor biotechnology firms. However, according to Steven Burrill, the majority of the big deals occur for late-stage products. Pharmaceuticals pay, on average, $115 million for phase III products, and the number of phase III deals has increased by 53 percent over 2001 (Burrill 2003). Late-stage products offer pharmaceuticals expensive but proven "quick fixes" for their ailing pipelines. One notable exception to this observation about late-stage deals is Roche, which typically prefers preclinical and phase I deals. In 2002, Roche struck two deals of over $200 million with companies that had early-phase products. According to Kevin Sachs and Gary Pinkus from McKinsey and Company, pharmaceuticals are overpaying for late-stage products to correct for their own pipeline deficiencies. The risks associated with earlier-stage products are offset by the value-creating deals. Furthermore, the additional risk can be diversified over a series of benchmarks.

Meanwhile, a few biotech companies that historically have been research-based are becoming more marketing/sales-orientated, and in some cases, they now compete with pharmaceutical firms. Examples include Amgen, Genentech, Biogen-IDEC, Genzyme, and Millennium. Biotechnology firms attempting to incorporate marketing early in the drug development process will face many challenges, including a major shift in company culture.

OPTIMIZING THE MARKETING FUNCTION

Based on our field study, we were able to extract the following general observations that help to maximize commercialization efficiencies.

PARTNERING AT AN EARLY STAGE OF DRUG DEVELOPMENT

Small biotechnology companies, specifically those without a commercial product but with a promising pipeline, can pursue different paths to profitability. Our field research suggests that one path is for a firm to obtain funding and mitigate excess risk by partnering with a big biotechnology or pharmaceutical company that has experience in bringing biologic products to market. Curis, a Boston-based

Minicase: Viagra—Market-Savvy Researchers Drive Success

There are many successful examples of marketing in the pharmaceutical field. However, none are so powerful as the case of Viagra.

Viagra is a blockbuster product for the pharmaceutical giant Pfizer. However, Viagra's path from discovery to commercialization was not straightforward. According to Pfizer's 1998 annual report, "Marketing of prescription pharmaceuticals depends to a degree on complex decisions about the scope of clinical trials made years before product approval" (Pfizer 1998). The history of Viagra clearly demonstrates that wise strategic decisions made by researchers during clinical trials can have a direct impact on the marketability of a drug. In particular, the success of Viagra can be attributed to the marketing acumen of the original scientists who recognized the potential for a drug that targeted patients with erectile dysfunction (ED).

The Road to Discovery

The original discovery team at Pfizer was led by Nicholas Terrett and Peter Ellis. In 1985, the team submitted a research proposal for a type 5 (PDE-5) inhibitor with possible application to hypertension. The registered chemical, UK92480, which is now known as Viagra, generated coronary artery activity and antithrombotic activity, enhanced nitric oxide, and had an antianginal effect. Based on these findings, Pfizer began clinical trials. In 1992, when Viagra was in its seventh clinical trial, the research team was focused on finding a drug that would treat angina patients who suffered from chest pain caused by blocked arteries. Unexpectedly, half of the volunteers in

biotechnology firm that focuses on regenerative medicine, is a good example of a company that pursues this path. Curis has various promising molecules at the preclinical stage and has already signed a codevelopment agreement with Genentech for one of these drugs, in addition to two other deals with Wyeth and Johnson and Johnson. These strategic alliances provide Curis with up-front cash and a share of future profits if the drugs reach the market.

one of the trial groups listed increased penile erection as a side effect. The researchers noted that Viagra was also increasing blood flow to the penis. Suddenly, what was little more than a side effect had potential as an application for treating erectile dysfunction. In 1994, Pfizer launched a trial with twelve patients who had ED and found the effects of Viagra to statistically improve their conditions. Viagra received approval from the FDA for ED on March 27, 1998.

The Erectile Dysfunction Market

Viagra revolutionized the way that ED is treated and was the first oral medicine for the condition. Prior to Viagra's launch, the most common method of treatment was an injection. Traditionally, urologists had treated patients with ED, but after Viagra was launched, primary care physicians became the main prescribers of the drug.

Marketing

The first week after Viagra was launched, in April 1998, an astounding 4.3 million prescriptions had been written. By the end of 1998, more than 200,000 doctors had written 7 million prescriptions for 50 million tablets, and Viagra was being sold in forty countries. Pfizer's pharmaceutical revenue in 1998 grew by 29 percent, with 12 percent of this growth attributed to Viagra, which had total sales of $788 million. Few drugs in history have attained such widespread use so quickly. Viagra has been called stage two of the sexual revolution (after the contraceptive pill) because of the effects it has had on society and attitudes toward sex.

The Viagra case highlights the importance of market understanding and indication identification in early-stage development. Pfizer was able to capitalize on a previously untapped market because of its marketing savvy.

Furthermore, they also increase the probability that these drugs will actually make it to market based on the marketing and approval process expertise that the partners bring to the table. However, to maximize how much Curis will receive in such strategic alliances, it must have an accurate picture of the potential value of its pipeline. This knowledge will help to increase its negotiating power and ensure that it is not being taken advantage of in the deal-making

Minicase: Enbrel—The Unexpected Winner

Enbrel provides another interesting case study on the role of marketing in the biotechnology industry. Immunex's failure to adequately conduct early market research precipitated a series of developments that culminated with the acquisition of Immunex by Amgen.

In November 1998, the FDA approved a treatment for rheumatoid arthritis (RA) that took a new approach—using a genetically engineered version of a natural body compound. The product was Immunex's Enbrel. Enbrel works by binding to and inactivating certain tumor necrosis factor (TNF) molecules before they can trigger inflammation. The RA market is estimated to comprise approximately 5 million patients in the economically developed regions of the world and 2 million patients in the United States alone. Though Enbrel was eventually approved for indications beyond RA, in addition to further indications within RA, this case study focuses only on the initial indication and the associated clinical studies.

Brief History

During the 1990s, a new class of drugs designed to attack the cause of rheumatoid arthritis, not just the symptoms, was being evaluated and researched. These drugs appeared to be effective without causing the usual side effects, such as stomach damage, associated with drugs available at the time for pain and inflammation. Clinical studies were started for some of the new drugs, including Enbrel.

In September 1997, Enbrel passed phase III trials, and Immunex was preparing to file for FDA approval sometime during the first half of 1998. Edward Fritzky, the chief executive officer of Immunex, stated in an interview in September that one-third of RA patients in the United States were not adequately controlled on their existing therapies and would therefore be ideal candidates for Enbrel (Fox News 1997). Later that month, Immunex signed a deal worth $100 million with American Home Products Corporation (AHP, now Wyeth) to market Enbrel for all indications outside oncology. Because all indications to date were unrelated to oncology, AHP would essentially be fully promoting Enbrel.

In March 1998, Fritzky had stated his belief that the overall potential of Enbrel would encompass approximately 1.8 million patients (Dow Jones News Service 1998a). This number was larger than initial predictions because Immunex now believed the drug would eventually be approved for earlier-stage RA patients as well.

Forecasting Phase

In July 1998, an analyst at Adams Harkness and Hill, Inc., projected a U.S. patient base of 400,000 to 500,000 patients for Enbrel (Dow Jones News Service 1998b). This projection reflected the number of persons for whom all traditional therapies had failed. The number was then expanded when the European market potential came on line. Sales of Enbrel, according to the analyst, were expected to be $139 million in 1999, $341 million in 2000, and $441 million in 2001. Actual sales in 1999 alone were $360 million.

The FDA granted approval to Enbrel in November 1998 as a last line of treatment. In the same month, Immunex announced its plan to file a supplemental biologics license application with the FDA by the end of the year to market the drug for children. Wyeth, AHP's pharmaceutical division, also filed an application with the European Market for approval of Enbrel in Europe. Based on Immunex's quarterly Securities and Exchange (SEC) filings, industry analysts stated that Immunex believed first-year sales of Enbrel would be greater than $250 million. These filings provided the first public indications that Immunex thought it had a better drug in its arsenal than industry experts had initially imagined.

The Good, the Bad, and the Ugly

The good news was that Enbrel surpassed nearly everyone's expectations. An interview with Fritzky in February 1999 provided some insight on why Enbrel became a blockbuster (CNBC/Dow Jones Business Video 1999). Fritzky attributed the initial explosion in sales to overwhelming approval of the drug from both patients and health care practitioners. He also stated that company executives were conservative about sales figures because they were concerned the initial figures might simply be a spike from trial users. This miscalculation would seem to underscore the basic thesis of this chapter. Immunex did not do an adequate job in researching its primary (or, for that matter, its total) market segment to determine potential adoption rates for Enbrel. By March 1999, 20,000 patients were signed up, and an additional 1,000 to 2,000 were signing up each week.

In May 1999, the FDA approved Enbrel for use in juvenile RA. This move essentially doubled the market potential, from 500,000 to 1 million patients. By this time, however, Immunex and AHP had realized the bad news—that their original sales expectations were grossly underestimated. To meet demand, additional facilities were dedicated to manufacturing Enbrel, but demand continued to exceed supply. By July 2003, thousands were on the waiting list for the drug, meaning the company sustained millions of dollars in lost revenue.

(continues)

What Went Wrong?

Immunex did not adequately perform its market research, which resulted in the underestimation of the Enbrel market. Though it seems Immunex correctly sized the market, it apparently overestimated the attrition rate, or the rate at which patients would try Enbrel but not continue using it. Immunex may have gained insight with a more rigorous marketing research campaign during the clinical trials. As it stands, AHP and Immunex did not realize Enbrel was a blockbuster until it was too late. And as a result, manufacturing constraints inhibited Enbrel's success for approximately three years. Had Immunex applied more marketing resources to its research efforts, the company's story might be very different today.

Hindsight Is 20/20

It is easy to point a finger at the lack of effective early-stage marketing/market development when a drug fails to meet expectations. But what happens when a drug under development is a breakthrough product with no analogue for comparison, as Enbrel was?

Forecasting market demand is a more complex task for a first-in-class, revolutionary product than for an evolutionary product. Enbrel, a first-in-class therapy for RA, exceeded the expectations of almost every expert in the field. According to Wyeth, it was the first biologic created to treat RA, and aside from Epogen and Procit, there were no other blockbuster biologic products on the market at the time. "The FDA was very concerned about immunologic problems (cancer as secondary effects) with Enbrel but safety proved to be higher than anticipated. Dealing with Enbrel was about dealing with the unknown, with no analogue for comparison" (Wyeth 2004).

Recognizing that it could not market the product alone, Immunex turned to Wyeth Pharmaceuticals for support. However, even with Wyeth's core competency in marketing, it was difficult to predict the uptake of Enbrel by rheumatologists.

The Enbrel life cycle has been a roller-coaster ride; facts, claims, studies, additional indications, and FDA approvals all played a part. Bringing in a marketing-based pharmaceutical company to help promote the drug and size the market is a potential solution, but this does not absolve a biotechnology firm for its failure to conduct in-depth marketing research. In Enbrel's case, it is evident that more comprehensive and earlier-stage market research at Immunex would have led to millions more in revenue for Immunex and its partner, Wyeth. And perhaps it would not have resulted in the acquisition of Immunex by Amgen.

Minicase: Natrecor—Better Late than Never

Scios's drug Natrecor is an example of a drug that was successfully launched and marketed by a biotechnology company. However, when the company was sold to Johnson and Johnson (JNJ) in 2003, the opportunities for the drug expanded because of the reach and the relationships of JNJ's sales force and the complementarities between Natrecor and certain existing JNJ products. Additionally, under the JNJ umbrella, Natrecor could be marketed in settings that Scios could not reach on its own. How much value could have been created and captured had Natrecor addressed these additional markets from the beginning? Although Scios had partnered with GlaxoSmithKline (GSK) for the European marketing rights to Natrecor, it did not have a big pharmaceutical partner to help it with the marketing of Natrecor in the United States until after the product was already on the market. Although it is difficult to quantify how much money Scios lost by delaying a partnership with a Pharma giant, it is not difficult to recognize that the synergies created by JNJ/Scios for the promotion of Natrecor could have been realized earlier had Scios partnered for U.S. sales prior to launch.

About Natrecor

Natrecor (nesiritide) is indicated for the treatment of acute congestive heart failure (CHF). The product, a recombinant B-type natriuretic peptide, is a synthetic human hormone that causes a vasodilatory response, and it represented the first approved treatment for acute CHF in nearly fifteen years. The FDA approved Natrecor for treatment of acute episodes of CHF on August 13, 2001, and the drug was launched in September 2001.

Scios's Marketing and Commercialization Plan

As of February 13, 2003, two and a half years after launch, Scios's marketing and commercialization efforts for Natrecor were still progressing. Approximately 88 percent of the targeted hospitals stocked Natrecor. Of those that stocked the drug in the third quarter of 2002, 87 percent reordered it in the fourth quarter. The Follow Up Serial Infusions of Nesiritide (FUSION) pilot study, which was evaluating the safety and feasibility of follow-up Natrecor serial infusions, was nearing completion. The Acute Decompensated Heart Failure National Registry (ADHERE) had enrolled over 35,000 patients. More than 350 clinicians and staff attended the second annual 2003 ADHERE national meeting in Palm Springs, California, earlier in

(continues)

2003. Enrollment continued in the longitudinal study module of the ADHERE registry that was set to follow patients with advanced heart failure for up to two years to assess changes in quality of life as a function of medical management and disease progression. Six abstracts related to Natrecor and acute congestive heart failure were scheduled to be presented at the 2003 American College of Cardiology annual meeting in Chicago (March 30 to April 2).

Later in February 2003, Scios announced that it was selling the company to JNJ; the acquisition was completed on April 29, 2003. Under the terms of the transaction, Scios shareholders would receive $45 for each outstanding Scios share. The transaction was valued at approximately $2.4 billion, net of cash.

Biotechnology/Pharmaceutical Synergies

At the time of the announcement, analysts believed that Natrecor could probably develop at a much faster rate with JNJ's institutional sales force and marketing efforts behind it. Morgan Stanley's analyst covering Scios at the time of the acquisition announcement estimated that U.S. sales of Natrecor in 2006 would be $336 million, assuming a successful result from Scios's FUSION trial, which was exploring earlier-stage use of the drug. It was assumed that JNJ had seen these data prior to the acquisition and thus considered them as a reason to move forward on this transaction. The Morgan Stanley analyst covering JNJ at the time of the announcement thought that JNJ could potentially increase sales of Natrecor to levels of $600 million in the United States in 2006, suggesting the synergies between JNJ and Scios could lead to an increase of approximately $264 million in Natrecor sales in 2006. Together, Scios and JNJ were considering new formulations with new administration alternatives using JNJ's Alza drug delivery technology.

The primary area of potential synergy was believed to exist in JNJ's extensive sales force and relationships in its Centocor and Ortho Biotechnology divisions. Natrecor was launched in September 2001, and Scios addressed the congestive heart failure opportunity through the clinical cardiologist. Centocor and Ortho Biotechnology would add their respective strong relationships in the emergency room and critical care unit—a potentially huge untapped market for Natrecor. Natrecor would complement the sale of Cypher, a sirolimus-eluting stent, and the Centecor drugs ReoPro (abciximab), for acute coronary care, and Retavase (retaplase), for heart attacks. JNJ and Scios also planned to expand Natrecor to the outpatient setting for less severe heart failure (based on results of Scios's FUSION phase II trial), as well as in hospital settings and emergency rooms.

In the case of Natrecor, Scios gained additional revenue from a product through synergies realized once it was purchased by JNJ. If Scios had identified these potential synergies earlier and done a more thorough assessment of its stand-alone ability to obtain sufficient reach for the product, it may have partnered (rather than sold) at an earlier stage and reached peak Natrecor sales sooner.

process. This assessment can only come through detailed, early-stage market research.

EARLY-STAGE MARKET DEVELOPMENT

Clearly, an efficacious marketing effort in the early stages of drug development is critical to the success of a biotechnology company. However, small biotechnology companies are science-driven and have limited financial resources, and thus, the marketing role is often left to the corporate development or business development arm of the company, if it is performed at all. Our field research suggests that in small biotechnology firms, functions are all interrelated and no marketing departments exist (Curis, Inc., and Immunogen, Inc. 2004). Typically, the strategic planning group conducts marketing for early-stage drugs. Yet some of these small companies have been able to incorporate robust marketing functions within their business development departments. Again, Curis is an apt example. "In the development process, it is extremely vital to determine early on which direction you put a product to market and into what indication. A successful product development effort must reflect commercial interests and consider reimbursement issues. Sometimes, biotechnology firms focus too much on science and can forget about the business part." (Missling 2004).

Two key elements are needed to establish an early development marketing function—people and attention. Specifically, qualified people must be found to conduct this activity, and senior management must devote attention to marketing and appreciate its importance.

People who will be effective in early development marketing should have a clear understanding of science as well as pharmaceutical/biotech marketing principles. Individuals who are comfortable dealing with uncertainty, thrive in unstructured environments, and can make qualitative assumptions are usually high performers. Despite the fact that "human beings are not used to making assumptions [and] tend to avoid them" (Missling 2004), in the biotechnology industry the earlier the assumptions are made for understanding the potential market, the better off the firm will be in terms of its ability to estimate the potential of a drug. The process for estimating a potential molecule market is an iterative process and therefore requires a person with good interpersonal skills to interact with many people, including scientists. "At Curis, the process to estimate a market is an iterative process, and I always involve people in order to obtain buy-in" (Missling 2004).

It is also vitally important to get senior management to "buy in" to the idea that marketing is critical. Management teams are usually too focused on funding issues, and they forget about other very significant functions of their companies. They also tend to believe that the drug benefits alone will be enough to successfully sell an approved product.

Genentech has successfully transitioned from a small biotechnology company to a leader in the field by both understanding the importance of marketing and cleverly integrating it with the science. Genentech still possesses a scientific culture because it allows its scientists to explore a wide array of drugs. However, the scientists are not completely free to explore any and every area that piques their interest. Instead, the marketing department has essentially identified key disease areas that have market potential, and the scientists are then allowed to explore molecules within these areas. This approach creates something of a checks-and-balances system between the R&D scientists and the marketing group. Furthermore, it gives the scientists freedom to innovate and the marketing group the assurance that the new developments will have viable market potential.

A FRAMEWORK FOR EFFECTIVE MARKET
PLANNING AND ANALYSIS

Strategic marketing and its incorporation throughout the drug development process is a key to the success of new product development at both biotechnology and pharmaceutical companies. Several critical marketing considerations should be examined well in advance of a product launch. To optimize the marketing efforts in the development of new biotechnology and pharmaceutical products, it is important for companies to examine these factors while their products are in development. Figure 8.1 provides some of these considerations at key points during the drug development process.

Although this model was designed for a large pharmaceutical company with a larger marketing staff than many small biotechnology companies have, it can still be effectively utilized on a more informal basis at smaller companies. Such companies may find this model useful in identifying important marketing-related considerations at various stages throughout the drug development process.

R&D	Preclinical	Phase I / IIa	Phase IIb / III	Launch
IP positions. Review discovery efforts to ensure alignment with strategic priorities. Disease research.	Vision and strategy for therapy. Prioritize compounds using decision tree models and considering competitive landscape. Identify key attributes. Develop product profiles. Refine understanding of doctor / patient dynamics and key value drivers.	Patient flow. Segments. Market entry strategy. Positioning platform. Clinical endpoints. Clincal sample population. Health economics. Pricing strategy. Phase IIa: Target key physicians, thought-leaders, and advocacy groups to increase awareness.	Geographic sequencing. Global positioning. Filing process. Publication plan. Key opinion leaders. Phase IIb: Develop pricing and reimbursement strategies.	New claims. New indications. New formulations. Channels. Global pricing.

Figure 8.1   Framework for Marketing Planning Process

Source: Personal communication between Johnson and Johnson executives and students.

Note: IP = intellectual property.

In the R&D phase, it is important to identify the intellectual property positions on the compounds and review the discovery
efforts to ensure they are in line with the overall strategic priorities of the company. Members of the marketing team at Curis are
very proactive about including themselves in the market planning
process as early as possible in the R&D phase. Because the company is smaller and has limited funds for the very expensive development process, it must carefully select the potential products to
pursue and then choose which ones to partner and which ones to
develop alone. To do this, the marketing team makes assumptions
and builds estimates of the potential U.S. markets for the products and indications that might be coming out of the R&D
department. It can then make more informed go-no-go decisions
and be more knowledgeable in terms of potential partnering and
alliance negotiations.

During the preclinical phase, it is important for companies to
begin developing a vision for the potential product as well as to
identify the key attributes and value drivers that will make the product succeed.

Once phase I clinical trials begin, the company should be able
to identify the minimum attributes that a compound must demonstrate in order to achieve success. During phase I/IIa trials, the
company should also start to examine the patient flow within the
market it hopes to enter, identify what clinical end points it will
eventually have to reach in order to effectively compete with products currently on the market, and begin to think about the potential economics and pricing of the product it is developing. During
phase IIa trials, the company may also want to begin targeting key
physicians, patient groups, and thought-leaders—people who will
be crucial in the development of the market for the drug (investigators, members of advocacy groups, key investors, and so forth—
to solicit important market research information and to increase
awareness and acceptance of the product the company is developing. After collecting this information, the company should be able
to make some informed management decisions regarding the clinical trial strategy going forward and garner more clarity about likely investment levels. The company should also have a solid under-

standing of the competitive landscape and what the positioning strategy of its product will be.

During the phase IIb/III stages of development, the company should be developing a publications plan, identifying and communicating with key opinion leaders, and finalizing pricing and reimbursement strategies. During filing with the FDA, the company should work to ensure that it has a competitive label and an appropriate channel strategy. After launch, the company needs to begin the process of life-cycle management and start to examine new claims, indications, and formulations for the product.

If companies are proactive about incorporating marketing earlier in the development cycle, they will be well prepared to launch their new products, as well as able to accelerate the time to peak sales. In biotechnology and pharmaceuticals, the latter capability is particularly important due to intellectual property concerns. Because a drug is covered by a patent for only a limited amount of time, companies should strive to attain peak sales as soon as possible to maximize their time at peak before patent expiration.

In summary, we would reiterate that biotechnology firms are often small organizations with limited resources and a focus on early-stage drug development. For these firms, the marketing effort is often overlooked. Based on our analysis, we believe that these firms will be better positioned over the long run if they allocate human resources to the marketing function at an early stage in the development cycle.

Some general and important reasons for incorporating this early-stage marketing can be summarized as follows:

- Market planning can contribute pertinent insights regarding go-no-go decisions, optimal indications, clinical outcome end points, and economic benefits.
- The market planning process gives information about potential end users and about the acceptance and adoption of new products. Similarly, clinical trial information that satisfies not only safety and efficacy concerns but also commercial endorsement needs is essential for maximizing profits. Safety and efficacy end

points help to guide decisionmakers in determining the potential levels of success for a product. If the expected outcome of a drug falls short of the end users' expectations, companies will avoid huge investments in further clinical trials.

- Market research allows the selection of optimal indications (see the Viagra case). Once the FDA accepts a drug for a certain therapy, that lead indication sets the tone for the entire product life cycle. Market size, competitive landscape, ease of data collection, and physician preference are a few important factors to consider when determining the most advantageous indications for new drugs. "Quantitative market research that analyzes physician preferences measured against factors such as price and efficacy can help the project team determine what the lead indication will be" (Garrett 2002). Market research, focused on future consumers, plays an important role in terms of clinical outcome end points. Recently, the FDA has begun to require an "outcomes" component for every NDA submission. This component typically consists of establishing significant differentiation for similar products. A clear example is the case of Viagra and Cialis, wherein the latter was the second to market a product with clear and significant clinical improvements in outcome end points. Thus, by incorporating the marketing research function earlier into the development process, a firm can increase its understanding of consumer needs and its ability to create clearly differentiated new products.
- Early marketing allows biotechnology companies to link economic benefits to clinical benefits. "The planning team must capture the appropriate pharmacoeconomic data in the clinical design of phase III clinical trials to generate economic benefit data that are consistent with managed care organization evaluation guidelines" (Garrett 2002). The objective of these firms should be

to reduce organizational risk, maximize success
potential, and avoid unprofitable investments.

Finally, we would note that the early incorporation of marketing into drug development offers multiple benefits. In the case of
Enbrel, Immunex underestimated the size of the market and lost
millions of potential dollars as a result—an outcome that may have
been averted had marketing been factored in at an early stage. Early-
stage marketing incorporation will allow small biotechnology companies to make more-informed decisions throughout the product
launch process. It may also position smaller biotechnology companies more favorably in negotiating strategic marketing alliances. In
the case of Natrecor, Scios gained additional revenue from a product through synergies realized once it was purchased by a larger
pharmaceutical company. If it had identified these potential synergies earlier and done a more thorough assessment of its stand-alone
ability to obtain sufficient reach for the product, it may have partnered at an earlier stage and reached peak Natrecor sales sooner. A
distinguishing factor of a market-oriented biotechnology firm is its
ability to recognize the potential value of early-stage preclinical
molecules and form deals with the right partner, at the right price,
and at the right time.

REFERENCES

Burrill, S. 2003. *Biotech 2003: 17th Annual Report on the Industry.* San
     Francisco: Burrill and Company, 333.
CNBC/Dow Jones Business Video. 1999. "Squawk Box—Immunex—
     Chairman & CEO—Interview," February 9.
Curis, Inc., and Immunogen, Inc. 2004. Roundtable discussion with Kellogg
     Biotech Ventures class during visit in Cambridge, Massachusetts,
     March.
Dow Jones News Service. 1998a. "Immunex CEO: Enbrel Could Help up to
     1.8M Patients," March 11. Available through subscription at
     http://www.dj.com.
———. 1998b. "Immunex: Adams Harkness Cites Arthritis Drug, Enbrel,"
     July 10. Available through subscription at http://www.dj.com.

Fox News. 1997. "Interview with the CEO of Immunex (Right after Phase III Study)." Cavato Business Report, September 12.

Garrett, S. 2002. "Eleven Steps to Successful Market Planning." *Pharmaceutical Executive,* special supplement (June), *Successful Product Management,* 6–18.

Missling, C. 2004. Presentation to Kellogg Biotech Ventures class during visit to Curis, Inc., March.

Pfizer. 1998. Annual Report. Available at http://www.pfizer.com/are/mn_investors.cfm.

Wyeth. 2004. Executive roundtable with Kellogg Biotech Ventures class during visit to Wyeth Biopharma, March.

Chapter 9

To DTC or Not to DTC: Direct-to-Consumer Marketing in Medical Devices

Roy Katz, Edna Lazar, Christopher Pfaff, and Thomas Walsh

M any competitive challenges in the life science industry have increased the burden on the marketing and sales functions of companies in this field today. At the same time, with additional distinct customers to satisfy and more "lifestyle" and personalized treatments, marketing and sales functions are more important than ever to the success of biotechnology and other life science firms. Total expenditures on prescription drug promotion by pharmaceutical companies grew about 70 percent from 1996 to 2000, and pharmaceutical companies now devote as much as 30 percent of their revenue to marketing and sales. Marketing medical products is a very complex undertaking and may be differentiated from marketing consumer products in three important ways: the type of customers, the regulatory constraints, and the value chain practices.

The marketing of traditional consumer goods must consider only two groups—the customers and the retailers—but pharmaceutical companies generally must take into account four groups—

patients, payers, doctors, and pharmacies. Marketing and sales efforts need to reach and provide benefits to each of these groups, which often have opposite needs and wants.

Adding to this complexity, drug marketing is a heavily regulated activity in the United States. The Food and Drug Administration (FDA) regulates all aspects of pharmaceutical marketing to ensure that health care professionals and the public are given balanced and truthful information about medicines and that all marketing claims are based on scientifically proven clinical evidence.

Traditionally, companies have considered the physician to be the end customer. However, direct-to-consumer advertising (DTC) is rapidly changing the established marketing landscape. In 1997, the FDA issued rules for television ads on prescription drugs. These rules loosened the restrictions on ads and encouraged pharmaceutical companies to target patients directly, which further increased the importance and complexity of the marketing and sales function within the pharmaceutical industry.

DTC must also meet federal requirements. An advertisement must mention the drug by name, list its side effects, and provide both contact information and a disclaimer regarding use of the drug by specific patients. Although DTC resembles traditional consumer marketing, the FDA has significant control over the advertisement content.

Television, the Internet, radio, and print media give drug marketers a variety of new channels to utilize. The Henry J. Kaiser Family Foundation found that every $1 spent on advertising in 2000 yielded an additional $4.2 in sales. Pharmaceutical advertisements targeting consumers now account for 15 percent of all U.S. drug advertisement spending, up from almost 9 percent in 1996. The biggest jump has occurred in TV commercials. The potential for Internet marketing is still not fully explored, but clearly, this medium offers a huge array of additional channels for expansion (Henry J. Kaiser Family Foundation 2003a).

DTC is shifting the decision power toward the patient and away from the doctor. Patients are now involved in their own diagnoses and choices of treatment. This situation presents an opportunity for biotech entrants to explore less traditional ways of marketing their products. But not all products are right for DTC. By watching TV, we can deduce that DTC is effective for therapeutic products. However, it is not yet clear whether DTC is effective for other areas in the life sciences like diagnostics or medical devices.

In this chapter, we provide a framework that will enable the marketing manager to determine if a product is an appropriate candidate for DTC marketing. As an example, we will use medical devices, which actually present a very interesting case because of their expanding market and their convergence with human therapeutics. So far, medical device companies have been reluctant to make use of direct-to-consumer marketing tools. As a benchmark for our case, we use the Big Pharma experience with DTC marketing.

THE CASE FOR DTC

We can easily argue that direct-to-consumer marketing increases commercialization efficiencies and offers major benefits to the public ("Pharmaceutical Advertising and You" 2003). These benefits are outlined in the following sections.

DTC INCREASES AWARENESS AND EDUCATES PATIENTS TO DISEASES AND THEIR TREATMENTS

Consumer advertising raises awareness of silent diseases (those with undetectable symptoms). And people remember drug campaigns: according to a recent survey, 81 percent of patients recall seeing direct-to-consumer ads for their conditions (PR Newswire 2003). In addition, people act in response to drug campaigns: 62 percent of patients between the ages of fifty and sixty-four said direct-to-consumer advertising has made them more aware of medication options they had not considered before. The same survey found two-thirds of Americans of all ages said these ads have allowed them to become more involved with their health care (McInturff 2001).

DTC DESTIGMATIZES DISEASES, WHICH IN TURN
INCREASES ACCEPTANCE

DTC is considered an excellent tool for destigmatizing conditions. A good example would be the campaigns for Viagra and Levitra, the erectile dysfunction drugs. Traditionally, of the 30 million American men estimated to have at least occasional problems getting and sustaining an erection, only about 10 percent seek treatment. Doctors report that a significantly higher number of men are talking about and seeking treatment for their erectile dysfunction conditions since TV commercials for these drugs began to appear (Allen 2003). However, the reactions of physicians to direct-to-consumer advertising are mixed. In a 1999 study by the FDA Division of Drug Marketing, Advertising and Communications, only 40 percent of the physician respondents said DTC had a very positive or somewhat positive impact on their patients and practices (Center for Drug Evaluation and Research 2004).

DTC INCREASES PATIENT COMPLIANCE

Often, one of the biggest issues in prescription medicine is that patients fail to take their medications or do not take them correctly. DTC advertising compels more patients to remain on medications and comply with the correct dosage and usage (Loden and Schooler 1998).

IS DTC A GOOD INVESTMENT?

In several surveys and studies, DTC has proven to have a bottom-line effect. A survey released by the Henry J. Kaiser Family Foundation showed that a 10 percent increase in direct-to-consumer spending increased drug sales by 1 percent (Henry J. Kaiser Family Foundation 2003b). According to a *New York Times* survey, about one-third of the 500 adults questioned in one month said they had asked their doctors about a medication they saw in a television commercial (Marino 2002). Of those who asked, 23 percent said they believed that they had the medical problem featured in the ad. What truly manifests the power of DTC is that 17 percent said that they would be willing to pay more for an advertised prescription drug even if it was not covered by their health insurance. It is evident that consumers are not only aware of DTC ads but are also *acting* on them.

The rise in spending on DTC by pharmaceutical companies is also evidence of their own belief that DTC is effective marketing. According to DataMonitor, the 2003 upward turn in DTC spending followed a two-year period of slow growth that saw DTC investment decline from a 44 percent increase in 2000 to a mere 8 percent rise in 2001, before leveling off in 2002. However, in the first six months of 2003, DTC spending reached $1.6 billion, an increase of 22 percent over the same period in 2002 ($1.3 billion). For the twelve-months ending June 2003—the most recent period analyzed in DataMonitor's 2003 study—DTC expenditures reached $2.8 billion, a 22 percent increase over the prior twelve-month period ($2.3 billion) (DataMonitor 2003).

DTC Works for Some but Not All

Direct-to-consumer advertising works better for therapeutics that treat chronic rather than acute diseases. It is also more effective with diseases that have clear and debilitating symptoms. Chronic diseases allow patients to spend more time researching and being attuned to their diseases. Drugs to treat high cholesterol, osteoporosis, or coronary artery disease are all good candidates for DTC. Consumers can evaluate their risk profiles and compare the available choices (DataMonitor 1998).

In addition, direct-to-consumer advertising also works well when the symptoms are easily recognizable and the patient's quality of life is affected by the disease. The most successful ads are those that treat allergies, stomach problems, and impotence—all of which have symptoms that are debilitating to the sufferer. In contrast, diseases such as high blood pressure and diabetes have symptoms that do not always interfere with daily life.

Investments in DTC advertising reflect these constraints. For the twelve-month period from July 2002 to June 2003, allergy drugs far exceeded any other category of drugs in DTC spending, with an impressive $549 million in DTC expenditures. Gastrointestial disorders generated $360 million in DTC investment, followed by asthma with $234 million. All of these drugs treat chronic diseases with debilitating symptoms (Lawrence 2003).

DTC can also work well for undertreated diseases in which education is important, as in the case of erectile dysfunction drugs. In addition, DTC is an effective tool for highlighting brand differentiation. For example, the Celebrex commercials are designed to make patients remember the brand name rather than to educate them about their condition.

NOT EVERYBODY AGREES THAT DTC IS GOOD FOR THE CONSUMER

Big Pharma is suffering from an increasingly negative public perception, mostly precipitated by pricing issues. The finger is pointed at the pharmaceutical industry's hefty profit margins. As prices have risen, so has the visibility of DTC advertising. The public remembers that, not so long ago, DTC was nonexistent and drug costs were lower. This correlation makes DTC a target. In fact, a National Consumers League survey from 2003 shows that almost half of all adults (49 percent) and 63 percent of seniors thought that drug ads are largely responsible for the increased cost of prescription drugs (National Consumers League 2003).

The higher visibility of DTC advertisements stimulated the FDA to become more watchful, and it demanded additional control of the marketing material used to advertise drugs. Recently, the FDA issued three draft guidance documents designed to improve communications to consumers and health care practitioners about medical conditions and products. The guidelines are the result of FDA research and policy development and were influenced by public participation at an open meeting on consumer-directed advertising (Lazarous 2003).

The criticisms boil down to a series of ethical questions. Will slick DTC promotions undermine the physician's authority? Will DTC create an artificial demand for inappropriate or overpriced treatments? Do advertisements overstate efficacy and understate risk? Are drug companies "creating" disease classes for conditions that were not recognized as diseases in the past—for example, attention deficit disorder (ADD) and attention deficit/hyperactivity disorder (ADHD)? At the end of the day, all advertisers, by definition, seek to have some effect on the consumer, and health care advertisers are no different. So the question remains whether it is wrong to "push" drugs to the end consumers or whether DTC helps consumers to have more control over their lives (White et al. 2004).

DTC Marketing and Medical Devices

Although DTC advertising has become a commonplace in the pharmaceutical industry, medical device manufacturers are acting more cautiously than most. Alongside advertisements for Propetia, Lipitor, and Viagra, one does not see ads for pacemakers, artificial knees, or stents. To this point, medical device consumer messaging has been largely limited to on-line sources, such as corporate websites and WebMD. This begs the question: how well does the pharmaceutical experience with DTC apply to medical devices ("U.S. Medical Supplies and Devices" 2004).

Compared to pharmaceuticals, medical devices treat a smaller, more specific set of ailments that can be discovered and treated via external electrical and mechanical means. Therapeutic devices treat both chronic and acute conditions. Vascular applications, particularly coronary ones, tend to treat more acute ailments. Orthopedic devices treat chronic conditions that will affect patients throughout their lives. Cardiac rhythm management devices (CRMs) can treat both chronic conditions (for example, bradyarrhythmia or congestive heart failure) and acute conditions (such as sudden cardiac arrest).

The diseases treated by therapeutic medical devices can be symptomatic (like most orthopedic illnesses) or asymptomatic (like most cardiovascular illnesses). Again, CRM therapies treat a mix of symptomatic and asymptomatic ailments. Congestive heart failure sufferers often have difficulty with physical activity, whereas patients who suffer from a sudden cardiac arrest rarely show symptoms before an acute event. Similarly, these devices treat life-threatening ailments (such as cardiovascular diseases) but also offer life-enhancing therapies (as with orthopedic ailments). Consumer choice is most prevalent for life-enhancing therapies, and thus, DTC advertising tends to be most effective in such markets.

Although pharmaceuticals are primarily administered either directly into the bloodstream or through oral means over a period of time or indefinitely, medical devices are almost always implanted permanently. For this reason, many medical devices are chosen at one point in time, and there is little likelihood that they will be replaced or renewed, limiting the opportunity to establish consumer loyalty. By contrast, many pharmaceutical therapies are ongoing, and prescriptions must be renewed; thus, consumer loyalty and compliance are key marketing drivers.

New drugs can compete if they are differentiated from existing drugs in small ways. Differentiation is an effective strategy because the drug's method of action is protected via patents and long development cycles. Within medical devices, the fundamental method of action is usually electrical or mechanical and is rarely unique to a particular manufacturer's device. Furthermore, because competitors can typically develop a competing product within a short period (one to three years), lasting differentiation becomes more difficult to achieve.

There are few differences in the customer profiles of pharmaceuticals and medical device manufacturers; both encompass physicians, payers, providers, and patients. Both types of companies use a common "push marketing" approach to gain physician brand preference. Payers often exercise more scrutiny with medical devices than with drugs. This reflects the fact that pharmaceuticals tend to be prescribed and paid for over a long period of time, during which recommendations can be adjusted and changed. By contrast, buying a medical device is often a "one-shot deal," and decisions made at that point are final.

Finally, the major difference between pharmaceuticals and medical device manufacturers in terms of customer base is their patient populations. Unlike that for blockbuster pharmaceuticals, the patient population for medical devices is small and nonrecurrent. Medical devices target specific problems that are often treated with one procedure. Therefore, the ongoing revenue potential for a particular therapy tends to be smaller, as are marketing budgets. With less money for marketing and a smaller target population to reach, medical device manufacturers historically have directed marketing resources toward physicians who could effectively reach their target market at a lower cost.

To DTC or Not to DTC: Decision Analysis

To determine whether DTC is an effective marketing strategy for medical devices, firms should follow a three-step process, consisting of an analysis of the competitive landscape, a consonance analysis, and a bottleneck assessment (Figure 9.1). Obviously, clear goals for the campaign also need to be determined early in the decisionmaking.

Before implementing a DTC program, it is important to specify what the strategy is expected to accomplish. In general, there are two broad goals: increasing awareness and education, on the one hand, and driving patient responses, on the other. Examples of each are found in the advertisements of pharmaceutical companies. Defining what the goals of the campaign are helps to facilitate the execution of the advertisement and the measurement of overall success. However, because the market for medical device advertising is in its infancy, we believe that the initial campaigns should be focused on augmenting awareness about the product and prompting specific patient action.

Increasing awareness is directly suited to this primary growth strategy. Interestingly, some firms have chosen to use an education campaign that does not employ the company name; only later, when the consumer is ready to pass from education to action, is he or she exposed to branding. At this stage, the consumer might be prompted to seek more information from a physician, to seek treatment directly, or to associate the product with the disease (brand recognition). Here, the importance of self-screening and channel preparation becomes clear. Good screening and preparation results in a match between a consumer—now a patient—and a physician who is ready to dispense the product.

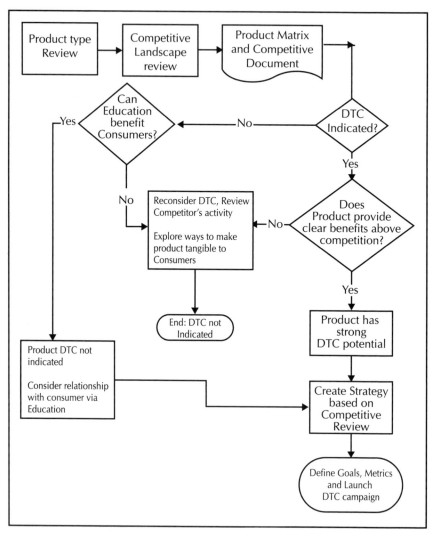

Figure 9.1 The Decision Process

ANALYZING THE COMPETITIVE LANDSCAPE

The competitive landscape analysis examines the product attributes, category and type, channel position, and brand position. Product characteristics and condition define the "value offering" of the medical device firm. Channel, brand, and competitive positions are concerned with the coordinates of the specific business.

Typical attributes of specific therapeutic medical devices are summarized in Table 9.1. Important considerations for determining the feasibility of a DTC campaign include symptoms, self-screening potential, and whether the disease is life-threatening. Ironically, consumers have minimal interaction with their implanted device. As opposed to pharmaceuticals, with each drug refill offering a new opportunity for the company to communicate with the patient, device manufacturers cannot use their devices as a means of communication. Once a device is implanted, the patient forgets about it. In addition, diseases treated by medical devices usually do not have recognizable symptoms, and the patient, unlike someone undergoing allergy treatment, cannot self-diagnose. More often than not, the patient being treated with a medical device has no time to consider different options; chances are he or she will rely on the physician's discretion, rather than any DTC campaign, in selecting a certain product. Table 9.2 presents a list of common disease attributes and indicates the feasibility of using DTC campaigns for each.

Table 9.1
Product Type Matrix

Typical Examples		Element of Product: Implant	Elements of the Condition		
Therapy Area	Example Product		Symptoms	Self-screening	Life-threatening
Orthopedics	Artificial knee	Yes	Yes	No	No
Cardiology	ICD	Yes	Yes	Yes	Yes
Cardiology	Stent	Yes	No	No	No
Neurology	Pain management	Yes	Yes	Yes	No

Note: ICD = implantable cardio defibrillator.

Table 9.2
Matrix Showing Common Disease Attributes and Important Questions to Consider to Determine Whether the Disease Is Conducive to Having Successful DTC Campaigns

Attribute	Questions to Consider*	Conducive to DTC	Not Conducive to DTC
Symptomatic	• Will the consumer recognize that something is wrong?	Yes	No
	• Does the condition present clear symptoms?	Yes	No
	• Are the symptoms chronic?	Yes	No
Self-diagnosis	• Can the consumer identify himself or herself as likely having a type of disease?	Yes	No
	• Is a patient able to recognize that the symptoms indicate a particular condition?	Yes	No
	• Can a score sheet be written to help the consumer self-diagnose?	Yes	No
Life-threatening	• Is the condition stable, with nonacute symptoms?	Yes	No
	• Does the condition first appear in a non-acute mode?	Yes	No
	• Will a symptomatic patient with this condition have time to consider options for treatment?	Yes	No
	• Can information be presented to consumers in an understandable format?	Yes	No
Product	• Does this product offer clear consumer benefits over its competitors?	Yes	No
	• Is this product the only existing product in the treatment category?	Yes	No

Notes: How well a disease "scores" for having a successful DTC campaign relates to how many "conducive" answers are recorded. Also, the manager must reflect on how difficult overcoming non-conducive elements will be.

* These questions assume equal weight among the considerations.

The channel and brand analysis focuses on understanding the motivations of each party in the channel (patients, providers, and payers), the power and influences that each party has over the others, and their possible reactions of the parties to DTC messages. Medical devices have a particularly complex channel distribution. The complexity arises from the delicate relationship between patients, providers (physicians and health care institutions), and payers (insurance companies and government agencies), whose goals and interests do not always align.

For medical devices that are purchased through an informed intermediary or implanted by a doctor, the relationship between patient and physician needs to be analyzed to estimate the possible effectiveness of a DTC campaign as well as potential problems, such as a backlash from doctors and/or consumers. By managing the relationship with the physician, medical device companies have an opportunity to prevent a buildup of tension like that that arose between the pharmaceutical companies and the intermediaries with the advent of DTC advertising.

There are two types of purchase decisions: the decision to purchase a device of a specific category and the decision to purchase a specific brand. In some cases, the decision is driven solely by the physician, who diagnoses the patient's clinical condition, determines that he or she needs a device, and then decides which brand to use. Other cases may be driven by a patient who has learned about a possible solution to a clinical condition that he or she has (or sometimes suspects) and then goes to the physician to request the use of the device and perhaps even a specific brand. In the case of drugs, we learned that, in many instances, patients received the prescription for the drug they requested. Between these two extremes, there is, of course, a complete set of decisions made together by patient and physician. The marketing manager needs to analyze the current decisionmaking process in regard to the product to be promoted. If the purchase decision for a particular product is driven by the patient, DTC seems to make a lot of sense because it does not change the existing doctor-patient relationship. But if the physician currently plays the main role in purchase decisionmaking for a given product, a DTC campaign would aim to change the current doctor-patient dynamics, potentially reflecting the relationship between the patient and the physician. In that case, the marketing manager needs to carefully assess possible push backs by physicians unhappy with having patients request specific treatments and even specific brands.

Even in cases in which the manufacturer is able to convince both the patient and the doctor that the device or brand is appropriate, other parties may actually have the final decision power. In some instances, hospital purchasing departments or insurance companies may restrict physicians from using specific devices. This situation is mainly true for devices that are not considered to be the "standard of care." In this circumstance, the marketing manager should judge whether a DTC campaign is powerful enough to create a strong demand from patients that will change organizational decisions regarding treatment and brand. Such an approach might be reasonable for life-saving devices that have proven to be significantly superior to the alternatives.

Finally, marketing managers should estimate the power of their product's brand. Is the brand strong enough to create a believable message? Part of believability has to do with how consonant the firm is with the consumer base. This factor is especially important when the campaign is trying to change the standard of care or physician and organizational preferences. If the DTC is successful in creating a pull demand, will this pull increase sales for the promoted brand? Or will other competing brands benefit from that marketing expenditure? Figure 9.2 summarizes the channels for the medical device industry as they exist today and as they will exist with future DTC.

The competitive position of therapeutic devices can be evaluated by identifying the leader in market share and determining how fierce the competitive response will be to the DTC campaign. Determining market leadership is not always a straightforward proposition. Markets are not always defined in clear lines, and products can be parts of other product portfolios. If a firm is dominant or alone in a product space, it has a clear advantage. However, the company still needs to consider whether the competition is concentrated or whether it consists of many firms vying for position. In the presence of concentrated competition, or duopoly, cooperation may be a suitable strategy for retaining a large portion of the market. A firm that secures its current market position is less likely to suffer a harsh response from the competition. Thus, competitive reaction and market position should be considered in tandem.

Market maturity affects growth opportunities and competitive behavior. If the market is immature, a firm that moves to isolate a niche is not likely to generate as forceful a response from its competitors as a firm in a mature market that attempts to dominate it. As markets mature, product differentiation becomes more critical to maintaining a profit margin. A special case in the pharmaceutical and biotech industry involves consumers who do not yet recognize that they need to be consumers—the undiagnosed segment. Thus, education is seen as a leading driver of market growth, as more people come

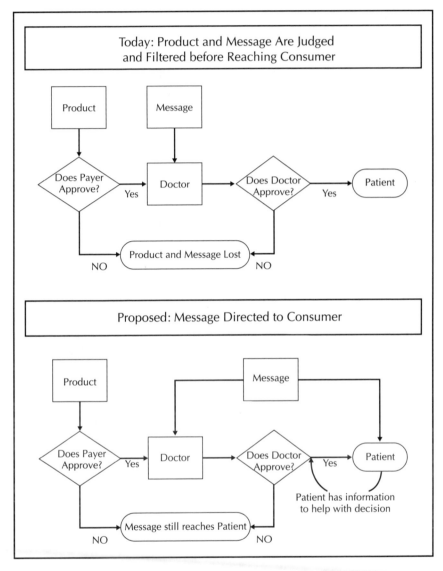

Figure 9.2   Knowledge Channels for Medical Devices Today and with Future DTC

to understand that they need treatment. Consider, for example, the campaigns of pharmaceutical companies to test bone density as a means to increase prescriptions for osteoporosis drugs. Market maturity can be thought of in terms of how well diagnosed the market is. In this context, we define "primary growth" as market growth attained via diagnoses and education and "secondary growth" as growth attained through the capture of more consumers. Primary growth helps the industry, and secondary growth helps an individual firm.

PERFORM A CONSONANCE ANALYSIS

Once the product's characteristics are defined, one can think of different paths to maximize its potential. Consonance analysis and core competence review seek to link the needs of the market with the capabilities of the firm. The boundaries of the defined market may need to be adjusted to best utilize the abilities of the firm. Two general questions to consider are: Does the firm have an advantage if the overall market grows? And does the firm have an advantage if a particular segment grows? Consonance and competence speak directly to these issues. If a firm chooses to pursue a strategy not consonant with its skills, it must dedicate itself to acquiring the needed skills. The manager must balance the cost of corporate learning with the expected benefits of the strategy. Two opposing strategies will emerge from the analysis: augmenting the overall market and concentrating on a specific niche (Figure 9.3). Although growing by capturing a larger market share requires that the product already be considered the market leader and that there are channels capable of absorbing high demand and overall education, competing in a small niche allows new entrants in the field to compete with sophistication in segmentation, targeting, and positioning capabilities.

FIND THE BOTTLENECKS

Three common bottlenecks in DTC campaigns have become the Achilles' heel of DTC: the channels; the public; and the segmentation, targeting, and positioning capabilities.

Channels
The product's channels must be ready to accommodate a DTC campaign. Channels, as the name suggests, can be blocked, too nar-

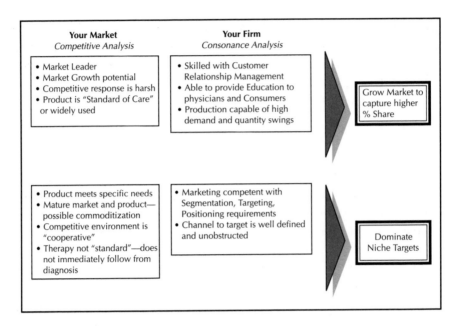

Figure 9.3   Competitive Landscape Assessment

row, or even altered. DTC has the potential to drive many customers toward a product offering. However, if the channel is not
prepared, then the DTC efforts can actually be detrimental.

DTC campaigns create a "pull" force that drives potential
patients to the providers. This pull can clog the channels by generating more patients than the channels can accommodate. This situation can be overcome by providing self-screening incentives. If
DTC is designed to encourage patients to screen themselves for
potential treatment, the campaign then generates important leads
while limiting unnecessary physician visits. Preeducating physicians
about the marketing efforts also improves relations by giving them
a sense of ownership in the DTC concept. The decision to launch
a DTC campaign first requires confirming that critical elements of
product delivery can be aligned with the firm's goals.

DTC should provide the consumer with the tools to interact
not only with the physician but also with the payers. A DTC campaign has the potential to rapidly increase a market and strain the

insurance companies. Any DTC strategy must include the payers' capacity to satisfy the increased demand. The manufacturer then needs to prepare and partner with all the elements of the channel network before launching a DTC campaign.

PUBLIC OPINION

DTC requires an exploration of public opinion concerns. Pharmaceutical firms, as noted previously, have come under fire for "wasting money" on DTC campaigns. Part of a DTC campaign involves the company or product image, and any threats to that image must be identified beforehand.

SEGMENTATION, TARGETING, AND POSITIONING

DTC is an expensive proposition, and we maintain that clearly defining the consumer will increase the likelihood of success, especially in the medical device industry where marketing budgets are far less extravagant than in the pharmaceutical industry. As such, segmentation, targeting, and positioning must be considered.

Demographic segmentation is particularly important in marketing medical devices because of the large underserved markets involved—people who have not yet recognized that they need treatment with medical devices. Underserved markets represent a primary growth opportunity that will be lost if exploited by others.

Television and magazines are the most common marketing venues; however, health care has additional access to customers through choke points that can bring in many potential customers very cheaply. A classic example is the waiting room. Doctors' waiting rooms, by their very nature, already screen out most of the population, and they represent an ideal location for advertisement placement. Choke points allow for the greatest exposure per dollar and should first be evaluated in the DTC campaign.

Clearly, the decision as to whether a DTC campaign should be pursued involves a complex analysis of the competitive landscape, consonance, and bottlenecks, as shown in the decision framework presented in Figure 9.3. To illustrate how this framework could be applied, we describe two cases using familiar products (see mini-cases). First, we show how the framework could work with Medtronic's implantable devices for pain management. Medtronic launched the "Tame the Pain" DTC marketing campaign to support these products. Next, we give an example of how the framework could apply to a different product that does not use a DTC campaign—Johnson and Johnson Cordis's Cypher drug-eluting stent.

Minicase: Medtronic "Tame the Pain"

Medtronic's Tame the Pain campaign was launched in November 2002. The purpose of this DTC campaign was to raise awareness of Medtronic's pain management products and induce patients to seek further information. Medtronic's Neurological and Diabetes Division manufactured medical devices that reduced pain by stimulating the patient's nerves with electrical impulses. The campaign, featuring Jerry Lewis as an advocate for the program in various media advertisements, increased awareness among potential patients about a possible treatment for chronic pain. Patients were also encouraged to log onto the program website, where they could seek further information. More important, they could take a self-assessment test that could tell them if the program's products might help them. Then the website would suggest specific specialist physicians affiliated with the program.

Competitive Landscape
In the category of implantable devices for pain management, Medtronic had almost complete control of the market. Competitors, to the extent they existed, controlled only a small fraction of the market. This suggests that for Medtronic, increasing the category was the same thing as increasing their own sales because there was no competitor that could steal this growth. Furthermore, there was a huge market potential for growth

(continues)

because there was a large population of both undiagnosed and undertreated patients. Many patients saw their primary care physicians (PCPs) for pain management, but many PCPs were not aware of the implantable pain management device option. The program's products, however, were not considered "standard of care," and therefore, reimbursement was either unavailable or difficult to attain.

Bottleneck Analysis

To see whether the product would have met the requirements for a DTC campaign, we should ask the questions shown in Table 9.3, where "yes" indicates a product is conducive to having a DTC campaign and "no" means it is not. From this table, we see there was a good fit between the product and a DTC campaign. There was a clear patient benefit that was very easy to explain in a nonacute chronic situation, and the consumer could play an important role in prescreening for the product.

Consonance Analysis

Medtronic had strong competence in providing information to physicians; however, it had less competence in directly contacting patients. To prepare for the Tame the Pain campaign launch, Medtronic was forced to develop various skills, including in messaging to patients and media selection.

Conclusion

The product had a very good fit for a DTC campaign, scoring "yes" answers across the DTC fit analysis. This means that Medtronic rightfully decided to promote pain management devices through the Tame the Pain DTC campaign. The only remaining question was how to go about implementing the campaign. This is where we revisit the competitive landscape analysis. Because Medtronic was a strong market leader in this category and because most of the potential market was not aware of the new pain management treatments, it was important to create a primary demand campaign, aimed at educating potential consumers. This was exactly what Medtronic hoped to accomplish in its Tame the Pain campaign: the company name was hardly mentioned, potential consumers were made aware of possible treatments, and they were encouraged to further investigate the product and later do self-screening. According to the self-screen results, patients could be

Table 9.3
Medtronic—Tame the Pain Bottleneck Analysis

Attributes	Questions to Consider	Tame the Pain*
Symptomatic	• Will the consumer recognize that something is wrong?	Yes (has pain)
	• Does the condition present clear symptoms?	Yes (pain)
	• Are the symptoms chronic	Yes
Self-selection diagnosis	• Can the consumer identify himself or herself as likely having a type of disease?	Yes
	• Is a patient able to recognize that the symptoms indicate a particular condition?	Yes
	• Can a score sheet be written to help the consumer self-diagnose?	Yes (can ask for location of pain, frequency, current medication taken and efficacy, etc.)
Life-threatening	• Does the condition first appear in a nonacute mode?	Yes (chronic pain is not acute)
	• Is the condition stable in this nonacute mode?	Yes
	• Will a symptomatic patient with this condition have time to consider options for treatment?	Yes (he or she can also try different products)
	• Can information be presented to consumers in an understandable format?	Yes (the product's operation and efficacy are straightforward)
Product	• Does this product offer clear consumer benefits over the competition?	Yes (the big competitors are drugs and for many patients, they do not offer real relief)
	• Is this product the only existing product in the treatment category?	Yes (this product is the only one in the pain management implantable devices category. Drugs are considered as substitutes, but this product is offered after drugs failed)

Score: # of "Yes" = 10   # of "No" = 0

offered the name of a specialist who could answer further questions. This approach was aligned with the providers' goals, as they received an increasing flow of prescreened patients. It was important that the website was successful in paring down the number of potential patients, so the channel (e.g., of specialists) was not clogged.

Minicase: Johnson and Johnson Cordis's Cypher Drug-Eluting Stent

Cypher is the first drug-eluting stent launched in the United States, and it was introduced by Johnson and Johnson Cordis in April 2003. A drug-eluting stent is a stent that includes a polymer impregnated with a drug that prevents cell growth. The drug significantly reduces rates of restenosis, or reblockage of the artery, from 35 percent to 3 percent. By thus reducing the need for additional angioplasty procedures, the drug-eluting stent creates significant value for patients, providers, and payers. Cypher was the only drug-eluting stent approved in the United States until March 2004, when Boston Scientific's Taxus stent was approved. Though at the time this chapter was written, it was too early to know which of the products would dominate the drug-eluting stent category, it is assumed by many analysts that Cypher will lose a large share of the market to Taxus. With this competitive situation, we would like to use the framework described in this chapter to determine if Johnson and Johnson Cordis should consider a DTC campaign for Cypher.

Competitive Landscape

Currently, Cypher and Taxus compete head to head in the drug-eluting stent category. It is difficult to determine which has the largest share, but we believe that it is safe to assume neither has a sustainable dominant share. In this situation, an attempt to increase the category might result in the competing product benefiting from the campaign. This suggests that any campaign considered should be focused on stealing share from the other product. However, there is still a potential for category growth. There are patients currently treated with bare-metal stents that might be converted to drug-eluting stents, and there are still a large number of underdiagnosed patients.

Both Cypher and Taxus therapies are reimbursed because they provide very high benefits to payers by reducing the need for additional angioplasty procedures. But at least currently, payers do not prefer one brand over the other.

There are differences between the products (different drugs, different delivery systems), but in the eyes of the patients, these differ-

ences are very subtle. Physicians might perceive a larger differentiation and might therefore strongly prefer one brand over the other.

Consonance Analysis

Johnson and Johnson Cordis is definitely skilled in physician education, but it has less experience in consumer education. The channel
to reach the patients is obstructed by the physician, who usually
makes the brand decision, especially because the product differentiation is more directed toward the physician—ease of use and
detailed specific medical testing. Johnson and Johnson Cordis suffered from manufacturing problems while launching Cypher, and it
might have problems with high swings in production levels caused
by a DTC campaign.

Bottleneck Analysis

Next, we test the fit between Cypher and a DTC campaign, again
using a "yes"/"no" DTC fit analysis. From Table 9.4, we see a bad fit
between DTC and the competitive situation Cypher faces. There is no
clear patient benefit (compared to Taxus), and many times, the product is needed in acute situations; further, consumers cannot self-
identify and screen themselves to determine if they need the product.

Conclusion

Currently, given its strong competition with Taxus, Cypher does
appear to be a good candidate for a DTC campaign. Because the
consumer value of both competing products is relatively equal, there
is no strong differentiation between the two in the eyes of the consumer. A DTC campaign to expand the category might be more successful because the drug-eluting stent category has a very high
patient value in comparison to bare-metal stents. But the problem
with a primary demand DTC campaign is that it may result in a large
gain for the competitor.

   The analysis shows that Johnson and Johnson Cordis should not
consider a DTC campaign at the present time. It is interesting to look
at the situation before Taxus was launched. At that point, Cypher was
the only product in the category, with significant patient value compared to the only alternative, bare-metal stents. In that situation,
Cypher would have scored five "yes" answers in the DTC fit analysis, which would make it a possible, if questionable, candidate for a
DTC campaign.

(continues)

Table 9.4
Johnson and Johnson Cordis Drug-Eluting Stent Bottleneck Analysis

Element	Questions to Consider	Cypher Drug-Eluting Stent*
Symptomatic	• Will the consumer recognize that something is wrong?	Yes. Patient might have chest pain, short breath before going to see the physician. But only after further testing will a diagnosis be made and use of a stent be recommended.
	• Does the condition present clear symptoms?	No, unless it becomes acute.
	• Are the symptoms chronic?	No.
Self-selection diagnosis	• Can the consumer identify himself or herself as likely having a type of disease?	No, unless the situation is acute.
	• Is a patient able to recognize that the symptoms indicate a particular condition?	No, he or she will need the assistance of a personal physician.
	• Can a score sheet be written to help the consumer self-diagnose?	No, the condition is not symptomatic in most nonacute stages.
Life-threatening	• Does the condition first appear in a nonacute mode?	No. Many times, the patient is diagnosed after an acute situation—a cardiac arrest, for example.
	• Is the condition stable in this nonacute mode?	Yes. If the condition is diagnosed in a nonacute mode, it is usually fairly stable in the short run (weeks).
	• Will a symptomatic patient with this condition have time to consider options for treatment?	Yes, but only if not in an acute situation.
	• Can information be presented to consumers in an understandable format?	No. Differences between products are too subtle.
Product	• Does this product offer clear consumer benefits over the competition?	No. The current research shows little difference in efficacy between Cypher and Taxus.
	• Is this product the only existing product in the treatment category?	No. Taxus is a direct competitor.

* Score: # of "Yes" = 3   # of "No" = 7

References

Allen, J. 2003. "Picking Up Where Viagra Left Off." *Health and Fitness.* September. Available at www.azcentral.com/health/men/articles/ 0911viagra-ON.html. Accessed June 2004.

Center for Drug Evaluation and Research. 2004. "Attitudes and Behaviors Associated with Direct-to-Consumer (DTC) Promotion and Prescription Drugs-Medical Division of Drug Marketing Advertising and Communication Survey." Food and Drug Administration. Available at http://www.fda.gov/cder/ddmac/dtcfllw.htm. Accessed June 2004.

DataMonitor. 1998. Press release, November 9. Available at http:// www.datamonitor.com.

———. 2003. "Pharmaceutical Campaign Management Synchronizing Online and Offline Consumer Marketing Strategies." Report #BFHC0595, June 24, 2003. Available at http://www.datamonitor .com. Accessed June 2004.

Henry J. Kaiser Family Foundation. 2003a. "Demand Effects of Recent Changes in Prescription Drug Promotions." Publication #6085, June. Available at www.kff.org. Accessed April 2004.

———. 2003b. "Impact of Direct to Consumer Advertisement on Prescription Drug Spending." Publication #6084, June. Available at www.kff.org. Accessed April 2004.

Lawrence, S. 2003. "Measuring Marketing." *Acumen* 1: 26–27.

Lazarous, David. 2003. *San Francisco Chronicle,* September 12.

Loden, D., and C. Schooler. 1998. "How to Make DTC Advertising Work Harder." *Medical Marketing & Media,* April. Available at http:// www.cpsnet.com/reprints/1998/04/DTCAdvertis.pdf. Accessed April 2004.

Marino, V. 2002. "All Those Commercials Payoff for Drug Makers." *New York Times,* February 24, 2002.

McInturff, B. 2001. "While Critics Might Fret, Public Likes DTC Ads." *Advertising Age,* March 26. Available at www.adage.com. Accessed April 2004.

National Consumers League. 2003. "Survey: Direct to Consumer Advertising of Prescription Drugs," January 9. Available at http://www.nclnet.org/dtcsurvey.htm. Accessed April 2004.

"Pharmaceutical Advertising and You." 2003. Health Partners Report, March.

PR Newswire. 2003. "DTC's Effectiveness in Driving Awareness: Doctor Contacts and Rx Requests Reach an All-Time High, According to New Market Measures/Cozint Research," October 28. Available at http://www.forrelease.com/D20031028/nytu141.P1.10282003123650.09079.html.

"U.S. Medical Supplies and Devices." 2004. UBS Investment Research Analyst Handbook, January 7.

White, H. J., L. P. Draves, R. Soong, and C. Moore. 2004. "'Ask Your Doctor!' Measuring the Effect of Direct-to-Consumer Communications in the World's Largest Healthcare Market." *International Journal of Advertising* 23: 53–68.

Chapter 10

THE FORGOTTEN ISSUE: REIMBURSEMENT IN BIOTECHNOLOGY

Raj Changrani, William Gangi, Alice Manard, Seung Sohn, and Steven Stark

Imagine the following scenario. A biotechnology company has just launched a new product to help fight a disease state typically found in patients with cancer. This product is believed to be best-in-class, with superior efficacy and longer-lasting effects than the current market leader. Significant resources and time were committed to its development because the company expects the product to enjoy wide and quick acceptance in the medical community and make immediate inroads on the incumbent product's market share. But six months after launch, sales in the most important patient segment begin to erode in one specific state. This state is not a high-volume state for the firm, but nevertheless, panic begins to set in as the company fears this trend might begin to hit other, more valuable states. The patient segment showing this drop in sales is comprised of individuals who depend on Medicare as their only form of medical insurance. The company has fallen victim to the local Medicare reimbursement policy, and as a result, it has lost millions of dollars in sales, with the potential to lose hundreds of millions more.

This scenario is fictitious, yet it is very similar to an experience that occurred during the recent launch of a blockbuster biologic from a well-known biotechnology company. The scenario highlights an important aspect of the biotechnology industry that is often forgotten—reimbursement.

THE CASE FOR REIMBURSEMENT

Understanding reimbursement and formulating a strategy to work with reimbursement are critical factors in achieving maximum sales potential for new products. Such products can have superior efficacy. They can have improved safety. They can be first-in-class therapies with elegant scientific formulas. But who is most likely to pay for the patient to use the product? This is the question physicians ask. Physicians use products they know will be paid for by some type of medical insurance. For many biotechnology products, that medical insurance is Medicare. Gaining coverage that allows for reimbursement at the highest level is a necessity to achieve maximum commercial success.

Biotechnology is a business of tremendous risk. In developing therapeutic products, there are numerous potential pitfalls. Biologics must be taken from preclinical development through a successful completion of phase III trials. The likelihood of success is small, and failure is possible at each step. Clinical trials must be developed correctly to reach desirable end points. Each day of delay in obtaining Food and Drug Administration (FDA) approval or tackling manufacturing challenges could mean the loss of millions of dollars in sales. All of these risks add up to an expected cost of $800 million for developing and commercializing a therapeutic product (DiMasi, Hansen, and Grabowski 2003). Reimbursement risk should not add to this development cost. Even if the biotech company is planning to outlicense the product to a Pharma, it should consider reimbursement as a de-risking tactic that maximizes the value of the product. With proper planning and strategy development, the biotech will be able to capture the greatest value during the licensing deal.

WHERE TO START: A QUICK
LANDSCAPE ANALYSIS

Analyzing the reimbursement status of a potential product is necessary through all stages of development. Understanding the competition and current reimbursement issues within the market helps identify sales potential. Sales forecasting and market potential are key to understanding whether a product has the potential to recover its developmental costs. Also, understanding the criteria for reimbursement eligibility and the various levels of reimbursement helps in devising the optimum design for clinical trials to meet specific outcomes and mitigate the effects of reimbursement barriers. Moreover, knowledge of the reimbursement mechanism and the current market is vital in the pricing strategy that is used for a product.

Biotechnology companies must develop a strong reimbursement analysis and management capability to examine and properly plan for these issues. This chapter introduces a framework to help companies determine reimbursement needs. In this framework, each company develops a "relative expertise in reimbursement" and "commercial maturity" value by scoring itself on several critical dimensions. Once these values are scored, the company can see how its reimbursement analysis and management capability compares to recommended industry benchmarks.

MEDICARE

The Centers for Medicare and Medicaid Services (CMS) is the federal entity under which Medicare falls. Medicare was established in 1965 to provide health care coverage to almost all Americans sixty-five years of age or older. In 1972, Medicare was expanded to include Americans living with disabilities and those suffering from end-stage renal disease (ESRD). Medicare is the single largest health insurance provider in the United States, covering approximately 15 percent of the population. Its expenditures in 2000 totaled $213 billion and accounted for 12 percent of all federal spending (Cann,

Rath, and Garett 2003). Medicare is divided into three parts (a fourth part will be added in 2006, encompassing the prescription benefits program). Part A is hospital insurance and helps pay for inpatient hospital services, skilled nursing facility services, home health services, and hospice care. Part B is medical insurance and helps pay for doctor services, outpatient hospital services, medical equipment and supplies, and other health services and supplies. Part C is Medicare+Choice and covers participant enrollment in health maintenance organizations (HMOs), preferred provider organizations (PPOs), provider sponsored organizations (PSOs), and medical savings accounts (MSAs). Part D, which will come on line in 2006, is the new prescription drug plan; it will help pay for retail prescriptions, which are currently not covered.

Biotechnology companies have a strong interest in Part B of Medicare, which supplies the reimbursement for therapeutic treatments. This reimbursement is of particular interest to biotechnology companies because of the high costs of biological drug development; without insurance coverage, many patients would not be able to afford therapy. A drug must meet various criteria to be eligible for reimbursement. In addition, a mechanism exists to describe the level of reimbursement and automate the process of providing payment to the physician. The following sections discuss reimbursement in terms of coverage, coding, and payment.

COVERAGE

Coverage refers to the treatments deemed reimbursable by Medicare. The two main dimensions to coverage are indication and dose administration. Indication refers to the disease or affliction that can be treated by a therapy according to FDA approval. Some products may have indications that are closely related; Amgen's Aranesp®, for example, is indicated for anemia due to chronic renal insufficiency in dialysis patients as well as anemia of cancer and chemotherapy-induced anemia in cancer patients. Indications also span across diseases, such as Genentech's Rituxan®, which is indicated for non-Hodgkin's lymphoma and is currently in clinical trials for rheumatoid arthritis. The reimbursement eligibility and level of payment for a new therapy are based on the specific indication for which the therapy is approved.

Medicare reimbursement is designed to lower the cost of receiving therapeutic treatment in a physician's office. Therefore, dose administration is the second key dimension to reimbursement. Medicare will only reimburse therapies that are injected subcutaneously, intramuscularly, or intravenously by the physician or are self-injected by the patient but require physician monitoring. These criteria impact the coverage of products in two ways. First, products that are designed with improved ease of use so that a patient can self-inject are not covered. Second, only acute therapies are covered. If a therapy is for a chronic condition, then the belief is that the therapy must have a patient self-administration capability. Thus, by definition, the therapy will not qualify for Medicare reimbursement. These two dimensions and several other stipulations are covered in an exhaustive list of requirements for reimbursement. For a therapy to be covered by Medicare, the therapy must satisfy the following criteria:

1. It must meet the definition of a drug or biological.
2. It cannot be self-administered.
3. It must be incident to a physician's services.
4. It must be reasonable and necessary for the diagnosis or treatment.
5. It cannot be excluded as an immunization.
6. It cannot be determined by the FDA to be less than effective.

Obtaining coverage by Medicare is the single highest priority in payment strategy for a drug. Because the government accounts for over 40 percent of national health care spending and Medicare is the most significant part of the federal program, Medicare is the market driver when it comes to determining health care reimbursement. Further, Medicare reimbursement decisions tend to set the standard for all payers in both the public and private sectors.

Medicare coverage decisions are made at both the national and local levels. A national coverage decision (NCD) is a statement granting, limiting, or excluding Medicare coverage for a specific medical service, procedure, or device. NCDs are binding for all Medicare contractors. Local coverage decisions (LCDs) are made by the local Medicare contractor, provided that no NCD exists. There are advantages and disadvantages to seeking reimbursement approval through both means. NCDs are all-encompassing and provide an economy of scale. LCDs take less time to establish, without the "all or none" risk involved with an NCD. However, LCDs involve a significant amount of work based on the number of local areas that need to be addressed.

CODING

Medicare reimburses drug costs through the Healthcare Common Procedure Coding System (HCPCS). In this system, a physician can seek reimbursement by submitting the code of the approved product that he or she administered to the patient. There are numerous codes within the HCPCS. Four main codes are of particular importance to biotechnology companies: the C-Code, J-Code, Q-Code, and S-Code. These codes are explained in Table 10.1.

The HCPCS codes are treatment-specific and not necessarily product-specific. This means that for a certain illness, a specific J-Code is used by the physician to gain reimbursement for the appropriate therapy employed to treat the illness. Multiple products from different biotechnology companies may fall under the same J-Code. The physician can choose among these products.

The J-Code is the preferred code because it is permanent. Physicians become comfortable with treatments and know that a therapy administered under a J-Code will be reimbursed. However, to obtain a J-Code, a company must have six months of marketing data and apply to the Centers for Medicare and Medicaid Services to obtain approval to use the code in April of each year. Therefore, based on the timing of a product launch, a company must decide whether to maintain a nonunique code prior to obtaining a J-Code or to use a Q-Code until a J-Code is obtained. Each method has its advantages. A Q-Code, similar to a J-Code, is considered a unique HCPCS code and therefore guarantees payment within forty-five days of product administration. However, physicians have their

Table 10.1
Health Care Common Procedure Coding System (HCPCS)

Drug Reimbursement Code	Code Type	Description
C-Code	Temporary	Used exclusively for services paid under the Outpatient Prospective Payment System and may not be used to bill services paid under other Medicare payment systems.
J-Code	Permanent	Used to report drugs given by injection. Drugs and biologics are usually covered by Medicare if: • They cannot be self-administered • They are not excluded (e.g., immunizations) • They are reasonable and necessary for the diagnosis or treatment of the illness or injury for which they are administered and they have not been determined by the FDA to be less than effective.
Q-Code	Temporary	CMS assigns Q-Codes to procedures, services, and supplies on a temporary basis. Q-Code can be replaced by a permanent code (i.e., J-Code).
S-Code	Temporary (non-Medicare)	Developed by commercial payer to report drugs, services, and supplies. They may not be used to bill services paid under any Medicare payment system.

office billing systems in place and become comfortable with the HCPCS Q-Code. Then, when a J-Code is issued, the physicians must take notice of the change and alter their systems as needed. A basic nonunique code does not entail the problem of adjusting automated billing systems or concerns about HCPCS coding changes. However, a basic nonunique code does not guarantee federal reimbursement within the forty-five-day period. The trade-offs between these two approaches must be analyzed to implement the optimal coding strategy.

PAYMENT

The final issue of Medicare reimbursement is payment. Once coverage has been approved and a code has been issued, the reimbursement level of a product must be determined.

As previously mentioned, codes are treatment-specific, so several products could be maintained within the same code. The government considers the average cost of generic products or of the least expensive branded products and then reimburses for the specific code at the lower of these levels.. For example, if both Product A and Product B, costing $150 and $175, respectively, are branded and fall under the same J-Code, the government will use the $150 price to determine the reimbursement level. If there are also generic drugs under the same J-Code with an average price of $75, then the price considered for reimbursement would be $75. This reimbursement scheme illustrates the importance of products showing enough difference from other products in terms of efficacy and safety to warrant the issuance of their own J-Code. In addition, for biologics, generics have not yet entered into the competitive landscape. If and when this does happen, the ability to maintain a unique J-Code will become even more critical.

The actual dollar amount reimbursed is determined by a pre-set algorithm. For 2004, the government reimbursed at 85 percent of the average wholesale price (AWP) (see the pricing description and example that follow [Blunt 2004]). With an 85 percent reimbursement and the patient copay, the physician achieves full cost recovery for administering the drug. In many cases, biotechnology companies give rebates to physicians, health care clinic franchises, group purchasing organizations, and the like so that the physicians make a profit in administering the product. Understanding the reimbursement mechanism, coding, and competitive landscape allows a biotechnology company to develop a pricing strategy to optimize sales and/or profit (Table 10.2).

Medicare closely monitors the reimbursement payments to lower the cost structure for physician office treatments. Three criteria are used to achieve this goal: the least costly alternative (LCA), functional equivalence (FE), and inherent reasonableness (IR).

In an LCA situation, Medicare designates two or more products as "clinically equivalent." Given this clinical equivalence, it reimburses for both products based on the AWP of the least expensive product. LCA decisions have been made for two drug classes to date—Lupron/Zoladex and Calcijex/Zemplar.

Table 10.2
Price Descriptions and Examples

Pricing Term	Definition of Term	Example Cost
AWP—average wholesale price	Average price of drug charged by wholesalers	$125
WAP—wholesale acquisition price	Price charged by manufacturer to the wholesalers	$100
ASP—average sales price	Average price physician pays after manufacturer rebates	$95
Medicare allowable	Amount for which Medicare deems Medicare and patients responsible (85% of AWP)	$106.25
Medicare reimbursement	Amount Medicare pays physician for administration of treatment (80% of Medicare allowable)	$85.00
Patient copay	Amount patient pays out of pocket for administration of treatment	$21.25
Physician profit	Amount physician receives above the cost to stock product	$11.25

An FE decision is based on the determination that two or more products function in the same manner. Once this determination is made, these products will be reimbursed at the lowest price of the two levels. An FE decision has been made only once, when the reimbursement of Aranesp® was lowered to the level of Procrit®. The Medicare Modernization Act (MMA) of 2003 prohibits further application of the functional equivalence policy.

The final tool is the IR. Inherent reasonableness allows Medicare, on a national or local basis, to reduce reimbursement for products by 15 percent per year if it is determined that the current reimbursement for a drug class is out of alignment with the reimbursement provided for similar products.

A FRAMEWORK FOR BIOTECH COMPANIES

Not unexpectedly, biotech companies vary dramatically in their ability to confront the challenges of reimbursement. Larger companies with multiple products on the market have entire departments devoted to ensuring maximum reimbursement levels on current products and those in the pipeline. Some even maintain a full-time staff in Washington, D.C., to monitor and affect policy decisions. Other companies have yet to launch a commercial product and thus have not dedicated resources to overcoming reimbursement hurdles.

The framework developed here attempts to identify opportunities for biotech companies to create value and increase the probability of commercial success by addressing reimbursement early in the product development process. Based on our preliminary research, we analyzed the relative expertise in reimbursement for a small sample of biotech companies of varying commercial maturity. We defined commercial maturity as a combination of current sales (in millions of dollars divided by one hundred) and the potential of the company's pipeline in terms of depth, quality, and stage of clinical development (potential is scored on a scale of 0 to 50). A company's relative expertise in reimbursement was calculated using a scoring system that incorporated the criteria in the following checklist:

- Has the company determined the administration mechanism for the product (injection, tablet, etc.)?
- Has it determined whether the product improves the efficacy or safety of existing treatment options?
- Has it determined whether partnering adds reimbursement expertise?
- Has it designed clinical trials to maximize reimbursement potential?
- Has it determined whether other forms of payment reimbursement are available for the product (e.g., special assistance for patients with a certain disease or affliction, such as AIDS)?
- Has it planned a coding strategy?
- Has it dedicated resources to ensuring reimbursement through local and national agencies?
- Has it hired lobbyists in Washington, D.C., to monitor and help drive policy?

Figure 10.1 plots a company's commercial maturity on the horizontal axis and its relative expertise in reimbursement along the

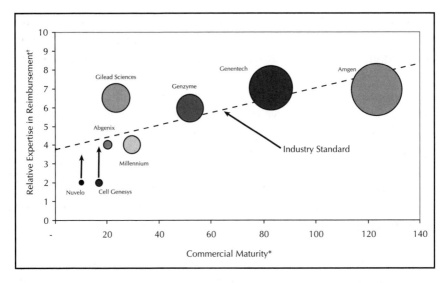

Figure 10.1   Relative Reimbursement Expertise for Select Biotechnology Companies

Sources: Data obtained from a variety of sources: company reports and interviews with company representatives.

Note: This chart plots a company's commercial maturity on the horizontal axis and its relative expertise in reimbursement along the vertical axis. The size of the bubble reflects the company's total enterprise value (TEV = market equity value + debt − excess cash). To adequately illustrate the properties of the smaller companies, the TEV values for Genentech and Amgen are halved.

* Commercial maturity reflects a combination of current sales (in $ millions divided by 100) and the potential of the company's pipeline in terms of depth, quality, and stage of clinical development (potential is scored on a scale of 0 to 50).

† Relative expertise is calculated by using a scoring system that considers: the administration mechanism for the product (injection, tablet, etc.); whether the product improves the efficacy or safety of existing treatment options; whether there are partners that add reimbursement expertise; the clinical trial design to maximize reimbursement potential; whether other forms of reimbursement are available (e.g., special assistance groups for patients with certain diseases or afflictions, such as AIDS); the plan coding strategy; whether resources are allocated for reimbursement and purposes; and whether the firm has lobbyists in Washington, D.C., to monitor and help drive policy.

vertical axis. The size of the bubble reflects the company's total enterprise value (TEV = market equity value + debt – excess cash). To adequately illustrate the properties of the smaller companies, the TEV values for Genentech and Amgen are halved.

Our analysis suggests that commercial maturity plays a large role in a company's relative expertise in reimbursement. The industry standard trend line shown in Figure 10.1 reflects where we believe a company should be in terms of reimbursement expertise for a given commercial maturity. Clearly, Genentech and Amgen have launched a number of successful products and are well versed in the process of ensuring maximum reimbursement for those products. Similarly, within the group of companies with a handful of commercialized products (Gilead Sciences, Genzyme, Millennium Pharmaceuticals), the ability to cope with issues of reimbursement appears to be fairly consistent. Although these companies could probably improve their in-house expertise somewhat, they are not substantially underemphasizing the importance of reimbursement.

More interestingly, there appears to be a fairly wide gap in the capacity for confronting reimbursement issues among companies that are in the earlier stages of commercial maturity. For example, Abgenix has not yet launched a commercial product but has partnered with Amgen for ABX-EGF (in late-stage clinical trials) and has partnered with AstraZeneca for up to thirty-six additional oncology products. Through these partnerships, Abgenix will most likely be able to leverage its partners' wealth of expertise when it is time to address reimbursement. Cell Genesys and Nuvelo are also in late-stage clinical trials with a few products but have yet to address the potential reimbursement issues to any meaningful extent. Based on our conversations with industry experts, we believe phase IIb in clinical development is an appropriate time to begin planning for reimbursement.

The therapeutic areas addressed by these companies' products (in oncology and cardiovascular diseases) represent an unmet medical need, which is likely to somewhat mitigate the risk of postponing the matter of addressing reimbursement. However, given the availability of third-party expertise to aid in planning for reimbursement, these companies have an opportunity to eliminate an

Minicase: MedImmune and FluMist

Roger Sampson, senior vice president of sales and marketing at MedImmune (Gaithersburg, Maryland), settled into his office. It was May 15, 2002, and Sampson needed to develop a strategy for positioning FluMist, MedImmune's new high-profile vaccine for influenza, in the influenza market. Sampson knew that the positioning strategy he developed for FluMist would have a major impact not only on the long-term sales potential for this drug but also on the long-term revenue growth of MedImmune.

MedImmune, Inc., is a global biotechnology company with 1,600 employees, 2002 revenues of $900 million, and an adjusted net income of more than $100 million. The company acquired California-based Aviron, Inc., in January 2002, for $1.3 billion in a tax-free stock-for-stock exchange in order to acquire its late-stage flu vaccine, FluMist, as well as several other drugs in various phases of clinical development. By the end of 2002, MedImmune marketed three different drugs. The company's lead product was Synagis (palivizumab), a humanized monoclonal antibody for use in the prevention of serious lower respiratory disease caused by respiratory syncytial virus (RSV) in pediatric patients at high risk of RSV disease. Synagis made up 79 percent of MedImmune's 2002 sales.

MedImmune also had a number of different drugs in development. The drug that was farthest along and that MedImmune had high hopes for was FluMist. Many industry analysts predicted that FluMist would reach $1 billion a year in annual sales.

Influenza

Influenza (also known as the flu) is the most widespread disease in the United States. Caused by the influenza virus, which is highly contagious and spreads through the air, the flu sends 114,000 people within the United States to the hospital each year. The flu affects the respiratory tract, including the nose, throat, and lungs, and produces symptoms that include fever, headache, tiredness, sore throat, cough, and aches. The flu is most common and most severe in high-risk groups (people over the age of sixty-five and under the age of five). These high-risk groups are most likely to experience the severe and often fatal side effects of the disease. Although the flu is mainly a life-threatening disease only for high-risk groups, it is still a

(continues)

major inconvenience for younger people who are healthy. Most people who have the flu stay in bed for several days to recover, missing work and other activities in the interim.

Flu Vaccines

The single best way to prevent the flu is to receive a flu vaccine. Flu vaccines help the body resist the influenza virus by building up the body's resistance. Flu vaccines were traditionally administered by injection and are still widely available. These flu shots are administered at hospitals, doctors' offices, pharmacies, workplaces, and elsewhere. Flu shots are relatively inexpensive, costing approximately $20 per shot, and many insurance companies cover the cost. The U.S. Centers for Disease Control (CDC) recommends that older adults, children, and people with particular health issues be vaccinated for the flu each year.

FluMist

The development of the FluMist vaccine represented a completely new way to provide a flu vaccine. For the first time, a flu vaccine could be delivered painlessly via nasal inhalation, rather than by an injection in the arm. MedImmune expected to receive approval to market FluMist from the U.S. Food and Drug Administration (FDA) in mid-2003. This would have allowed the company to market the drug during the 2003–2004 flu season. MedImmune anticipated that the FDA would allow FluMist to be used in healthy children and adults between the ages of five and forty-nine—a large population with approximately 160 million potential customers (see Table10.3). The CDC has set a 2010 goal of 120 to 150 million inoculations. The CDC has prioritized influenza as a major concern, and this is driving growth through increased awareness.

FluMist Positioning and Launch Strategy

Once FluMist was approved by the FDA, MedImmune needed to quickly get its marketing program up and running for the launch. The task of determining how to best position FluMist on launch was left to Roger Sampson. Once the positioning was set, it was relatively straightforward to determine the pricing, advertising strategy, marketing strategy, and so on. FluMist had the largest potential of any product MedImmune had developed, so Sampson knew a successful launch was essential.

Cost was a significant issue to factor into the positioning strategy. Manufacturing FluMist was a challenge, producing the vaccine

Table 10.3
Population Addressable by FluMist

Age Range	Population	Percent Healthy	Target Population
5–17	57,000,000	90	51,000,000
18–49	127,000,000	85	108,000,000
Total	184,000,000	86	159,000,000

was costly, and capacity was limited. For 2003, MedImmune could produce up to 5 million doses of the vaccine. At that production level, it cost $52 to produce each dose. However, MedImmune planned to ramp up production over the next several years, thus potentially lowering per-unit manufacturing costs.

Launch Plan

Sampson had asked the advertising agency Saatchi and Saatchi to develop a recommendation on how to position and launch MedImmune. The agency's recommendation was to: (1) position FluMist as a way to stay healthy; (2) price FluMist high; and (3) market it with a strong direct-to-consumer campaign, using consumer demand to pull the product through distribution channels. Sampson thought these recommendations made sense, and he was inclined to go with them as proposed, but there was considerable disagreement on his team. In particular, Susan Johanse, the vice president of financial planning, argued that there was no need for a strong DTC campaign. Robert Samuels, the product manager on FluMist, believed the best approach was to position FluMist as a way to avoid a shot. And Sing Song, the marketing strategist, suggested that they should determine their launch strategy after doing some preliminary market research to gauge consumer willingness to pay because consumers purchasing FluMist were going to have a higher copay or possibly receive no reimbursement at all.

Introducing FluMist

Based on input from the finance team, the product manager, and the marketing team, Roger Sampson decided to introduce FluMist at $49 and position it as "insurance against the flu." The team believed that the product was differentiated enough that when supported by the

(continues)

aggressive $20 million DTC campaign, the perceived benefits would be higher than $50 in the minds of the target consumers.

Results of the FluMist Launch

By the end of the first full flu season, it was apparent that the launch of FluMist was a major disappointment. Sales were nowhere near company or analyst predictions, and it appeared that the DTC campaign was inefficient and caused confusion. As a result of the disappointing launch, MedImmune's market capitalization fell by $3.5 billion, and its partner, Wyeth, was having second thoughts regarding its comarketing agreement with MedImmune for FluMist.

The main reason for poor sales was the unwillingness of consumers to pay $49 for FluMist as opposed to $20 for the alternative, the flu shot. It is quite possible that consumers would have been more eager to pay this 150 percent premium if MedImmune had only lobbied the right channels and obtained reimbursement for FluMist.

In fact, MedImmune failed to take action along several fronts with regard to the launch of FluMist. First, as Sing Song originally suggested, the company could have performed some up-front market research to consider the price elasticity of FluMist. Perhaps it would have found early on that consumers would be unwilling to pay a 150 percent premium for FluMist. Second, MedImmune could have analyzed how to obtain Medicare reimbursement for FluMist. Because Medicare does not provide reimbursement for treatments administered through nasal inhalation, MedImmune would have been facing an uphill battle. But perhaps it could have made a case to Medicare that by using a live vaccine in FluMist, the potential for increased efficacy existed, which might have prevailed over Medicare's rules regarding nasal administration. Finally, the company could have improved its market research to determine if the market potential of FluMist outweighed the costs of development (or, in this case, the purchasing of Aviron) and manufacturing the drug. Perhaps it would have realized at any early stage that the product was much too expensive to produce in order to get a decent return on its investment. In hindsight, by focusing on reimbursement early in the drug development process, MedImmune might have been able to: (1) gain coverage for FluMist, (2) alter its marketing strategy to position FluMist differently, or (3) negotiate a lower price for acquiring Aviron after reviewing market research data related to the potential pricing of FluMist.

unnecessary risk and increase shareholder value by addressing some of the items on the reimbursement checklist. For a company that is several years away from launching its first product, investing substantial resources to developing in-house expertise to address reimbursement is not a viable option. As an alternative to developing in-house expertise, a viable alternative for such a company may be to retain the services of a reimbursement-savvy consulting firm that specializes in commercialization strategies for early-stage products. Alternatively, partnerships can be valuable resources from a reimbursement perspective, as they provide an early-stage company with an opportunity to gain insight into the reimbursement process from its more experienced partners.

Reimbursement is commonly forgotten in the day-to-day operations of a biotech company. Nevertheless, formulating a reimbursement strategy early on is critical to the successful launch of a new product. Biotechnology companies must be aware of Medicare requirements and begin developing their strategies in the early stages of drug development. Most industry experts agree that by phase IIb, a biotechnology company should have already taken the initial steps of gaining reimbursement for its product. At a minimum, company managers should understand: (1) the current reimbursement levels for competing products, (2) the criteria for reimbursement eligibility and the various levels of reimbursement, and (3) the reimbursement mechanisms and the current market. By understanding these issues early on, a company will be able to mitigate its risks by optimizing the design of clinical trials, performing more accurate market research, and developing better sales forecasts. Most important, this knowledge and preparation will ensure the company obtains reimbursement at the highest level possible, thus maximizing commercial success.

REFERENCES

Blunt, Roshawn. 2004. Reimbursement Lecture. Kellogg School of Management, February 25.

Cann, Leah, Vanessa Rath, and David Garett. 2003. "A Reimbursement Handbook for Drugs and Biologics." Wachovia Securities, May.

DiMasi, Joseph, Ronald Hansen, and Henry Grabowski. 2003. "The Price of Innovation: New Estimates of Drug Development Costs." *Journal of Health Economics* 22: 151–185.

INDEX

About the Editor

Dr. Alicia Löffler is director and professor of biotechnology at the Northwestern University Kellogg Center for Biotechnology at the Kellogg School of Management, Evanston, Illinois. She is in charge of designing and running all the biotechnology, pharmaceutical, and life science activities including the MBA biotech program, executive programs, and research activities. Previously, she directed and ran the Northwestern University Center for Biotechnology (Office of Vice President for Research). Dr. Löffler created the center's educational programs including the master's program in biotechnology, the summer Biotechnology Institute, and career development programs. She is globally recognized as a pioneer and leader in biotechnology education and speaks and consults widely with firms and universities in the United States, Asia, and Europe.

Dr. Löffler actively serves as adviser to many external organizations including: Baird Venture Partners; founding board member of the Biotechnology Institute, Washington, D.C.; Biopharmaceutical Center at the WHU in Koblenz, Germany; Midwest Heart at Edward Hospital (Illinois); Women Entrepreneurial Center (Illinois); and Industrial Management and Data Systems of the Emerald Group, London. She also served as board member and past chair, Council for Biotechnology Centers (BIO), and board member, Emerging Companies, Biotechnology Industry Organization, and the Governor's Edgar Council for Biotechnology. She was recently named one of the tech one hundred stars by Crain's *Chicago Business.*

Dr. Löffler received her bachelor of science from the University of Minnesota, her doctorate from the University of Massachusetts, and did postdoctoral work in biochemical engineering at Caltech, Pasadena, California.